Advances in Numerical Mathematics

Gerhard Zumbusch

Parallel Multilevel Methods

Advances in Numerical Mathematics

Editors

Hans Georg Bock
Wolfgang Hackbusch
Mitchell Luskin
Rolf Rannacher

Gerhard Zumbusch

Parallel Multilevel Methods

Adaptive Mesh Refinement and Loadbalancing

Teubner

B. G. Teubner Stuttgart · Leipzig · Wiesbaden

Bibliografische Information der Deutschen Bibliothek
Die Deutsche Bibliothek verzeichnet diese Publikation in der Deutschen Nationalbibliographie;
detaillierte bibliografische Daten sind im Internet über <http://dnb.ddb.de> abrufbar.

Prof. Dr. Gerhard Zumbusch
Geboren 1968 in Münster. Studium der Mathematik 1987-1992 an der TU München, Diplom. Von
1993 bis 1995 Konrad-Zuse-Zentrum für Informationstechnik Berlin, Promotion 1995 FU Berlin, an-
schließend SINTEF Anvendt Matematikk Oslo 1996. Danach Universität Bonn 1997-2002, Habilitation
2001, Privat-Dozent 2002. Seit 2002 Professor an der Friedrich-Schiller-Universität Jena, Lehrstuhl
für Wissenschaftliches Rechnen/Numerische Mathematik, Direktor des Instituts für Angewandte
Mathematik.

1. Auflage November 2003

Der B. G. Teubner Verlag ist ein Unternehmen von Springer Science+Business Media.
www.teubner.de

Umschlaggestaltung: Ulrike Weigel, www.CorporateDesignGroup.de

Gedruckt auf säurefreiem und chlorfrei gebleichtem Papier.

ISBN-13:978-3-519-00451-6   e-ISBN-13:978-3-322-80063-3
DOI: 10.1007/978-3-322-80063-3

Preface

Numerical simulation promises new insight in science and engineering. In addition to the traditional ways to perform research in science, that is laboratory experiments and theoretical work, a third way is being established: *numerical simulation*. It is based on both mathematical models and experiments conducted on a computer. The discipline of scientific computing combines all aspects of numerical simulation. The typical approach in scientific computing includes modelling, numerics and simulation, see Figure 1.

Quite a lot of phenomena in science and engineering can be modelled by partial differential equations (PDEs). In order to produce accurate results, complex models and high resolution simulations are needed. While it is easy to increase the precision of a simulation, the computational cost of doing so is often prohibitive. Highly efficient simulation methods are needed to overcome this problem. This includes three building blocks for computational efficiency, discretisation, solver and computer.

Adaptive mesh refinement, high order and sparse grid methods lead to discretisations of partial differential equations with a low number of degrees of freedom. Multilevel iterative solvers decrease the amount of work per degree of freedom for the solution of discretised equation systems. Massively parallel computers increase the computational power available for a single simulation. However, parallel computers require parallel algorithms and special methods to code them including data distribution and communication, which poses a severe problem for adaptive mesh refinement. Furthermore multilevel solvers have to be specifically tailored so that they can be applied to the adaptive discretisation. Even the efficient implementation of multilevel methods for sequential and parallel computers poses a severe problem. These aspects will be covered in detail in the following chapters.

Last but not least, let me thank all who supported the present work in one way or another. To name but a few, let me begin with my supervisor Prof. M. Griebel, who supported my research over many years and created a research environment which was probably unique at a mathematics department.

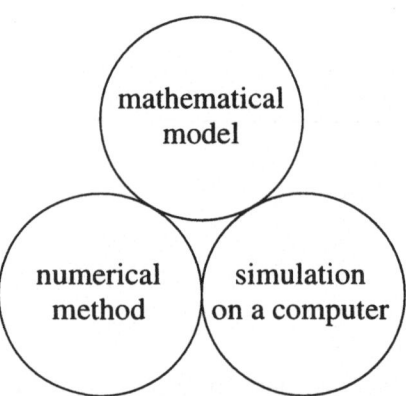

Figure 1. *Three ingredients of scientific computing: a mathematical model, a numerical method and the simulation on a computer.*

This enabled the combination of ideas from fields as diverse as approximation theory and molecular dynamics, multilevel methods and high speed networking. Furthermore, the leading edge equipment allowed for many projects years before it became close being mainstream. However, he also contributed the basic idea of the present work, namely the idea of applying space-filling curve techniques from astrophysical particle methods to parallel adaptive multigrid methods, a topic I worked on ten years ago at TU München then with his support and supervised by Prof. R. Hoppe.

Of course I have to thank the whole group *Scientific Computing and Numerical Simulation*, members of the *Institute for Applied Mathematics* and members of the SFB 256 (Sonderforschungsbereich) *Non-linear Partial Differential Equations*. Let me name some of them individually, e.g. M. A. Schweitzer for the collaboration on the construction of our cluster computing resources and discussions on multigrid methods and parallelisation in general. The sparse grid and wavelet parts were influenced by F. Koster and T. Schiekofer, who calculated some wavelet coefficients for the best approximation results and laid the algorithmic foundations of the finite difference sparse grid discretisation respectively. Some research related to space-filling curves was done by M. Ellerbrake and G. Spahn, who created the pictures of the tetrahedron meshes. The calculations on the T3E at Cray Inc. were supervised by M. Arndt. Furthermore I want to thank Prof. P. Oswald (Lucent) and Prof. H.-J. Bungartz (Stuttgart) for useful discussions on sparse grids and space-filling curves respectively. P. Anderson, M. Arndt, M. Bader, F. Kiefer and M. A. Schweitzer did some proof reading. Thanks also to the referees of the original thesis text and the editors of the book series for their effort of reviewing, their patience, and their comments. I also wish to express my gratitude to Teubner-Verlag for their friendly cooperation.

Finally I have to thank SFB 256 at *Universität Bonn* for the financial support, the *Institute for Scientific Computing Research* (ISCR) and members of the *Center for Applied Scientific Computing* (CASC) at *Lawrence Livermore National Laboratory* for the opportunity to stay there as a guest and to access their computer resources, namely the ASCI Blue Pacific computer and several smaller systems. Furthermore I have to thank Cray Inc. and NIC (*Forschungszentrum Jülich*) for the computing time on their Cray T3E systems.

Jena, August 2003   *Gerhard Zumbusch*

Contents

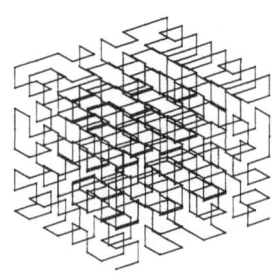

Chapter 1

Introduction

In a short example we want to illustrate some of the concepts this book is about. Let us consider the two dimensional Poisson problem as a homogeneous Dirichlet boundary value problem

$$
\begin{aligned}
-\Delta u &= f \quad \text{in } \Omega, \\
u &= 0 \quad \text{on } \partial\Omega,
\end{aligned}
\tag{1.1}
$$

where $\Delta u = \partial^2 u/\partial x^2 + \partial^2 u/\partial y^2$ is the Laplace operator, Ω is a bounded, open domain like $[0,1]^2$ whose boundary is dentoed by $\partial\Omega$. We are looking for a solution u as a function $\Omega \mapsto \mathbb{R}$ for a given right hand side function $f : \Omega \mapsto \mathbb{R}$. The finite difference approximation of the problem is based on a discretisation of u at grid points $x_{i,j}$ defined by $x_{i,j} = (ih, jh)$ with $0 \le i,j \le N$ and mesh size $h = 1/N$. We denote the discretised solution u_h at the grid pionts by $u_{i,j} = u_h(x_{i,j})$, the right hand side accordingly by $f_{i,j} = f(x_{i,j})$ and write the finite difference stencil as

$$
\begin{aligned}
\tfrac{1}{h^2}\left(4u_{i,j} - u_{i+1,j} - u_{i-1,j} - u_{i,j+1} - u_{i,j-1}\right) &= f_{i,j} \quad \text{for } 0 < i,j < N, \\
u_{i,j} &= 0 \quad \text{else.}
\end{aligned}
\tag{1.2}
$$

Putting the Taylor expansion of u_h like

$$
\begin{aligned}
u_h(x_{i\pm1,j}) = u_h(x_{i,j}) \ &\pm\ h\tfrac{\partial}{\partial x_1}u_h(x_{i,j}) + \tfrac{h^2}{2}\tfrac{\partial^2}{\partial x_1^2}u_h(x_{i,j}) \\
&\pm\ \tfrac{h^3}{6}\tfrac{\partial^3}{\partial x_1^3}u_h(x_{i,j}) + \tfrac{h^4}{24}\tfrac{\partial^4}{\partial x_1^4}u_h(\zeta)
\end{aligned}
$$

with some point ζ into (1.2) it turns out that the centered finite difference stencil does indeed approximate the Laplace operator second order accurate for sufficiently smooth u and u_h. The conditions (1.2) can be rewritten as an equation system

$$
A_h u_h = f_h
\tag{1.3}
$$

with vectors u_h, $f_h \in \mathbb{R}^n$, matrix $A_h = (a_{k,l}) \in \mathbb{R}^{n \times n}$, and $n = (N-1)^2$. The matrix entries with an enumeration of the unknowns by $k = i + j(N-1)$ are given by

$$a_{k,l} = \frac{1}{h^2} \begin{cases} 4 & \text{if} \quad k = l, \\ -1 & \text{if} \quad k - l = 1, -1, N-1, \text{ or } -N+1, \\ 0 & \text{else.} \end{cases}$$

The matrix A_h is sparse in the sense that it contains only $\mathcal{O}(n)$ non-zero entries. However, the solution of the equation system (1.3) by standard Gaussian elimination requires $\mathcal{O}(n^3)$ arithmetic operations.

We conclude that a solution of accuracy $\epsilon = \mathcal{O}(h^2) = \mathcal{O}(N^{-2}) = \mathcal{O}(1/n)$ requires $\mathcal{O}(n^3) = \mathcal{O}(N^6)$ arithmetic operations. This means roughly eight-fold operations in order to reduce the approximation error be one half. For this simple example there are several way to improve this ratio of work to accuracy.

First of all, we can use solvers of the equation system (1.3) which exploit the structure of the matrix A_h or even properties of the solution u. This leads us to multilevel and multigrid methods which reduce the arithmetic operations to $\mathcal{O}(n)$. Next we can improve the discretisation scheme (1.2). Higher order schemes usually show higher accuracy than our second order central differences. However, in the presence of non-smooth solutions u we do not even get $\epsilon = \mathcal{O}(1/n)$ and we can turn to adaptive mesh refinement. The goal is to minimise the number of unknowns n for a given accuracy ϵ and a given problem. Another concept are sparse grids, where one reduces n for a given accuracy ϵ a priori. As a third way to reduce computing time we can use parallel computers. The arithmetic operations are distributed to several independently operating processors such that the time to perform n operations reduces to values below $\mathcal{O}(n)$. Using p processors some parallel algorithms require only $\mathcal{O}(n/p + \log p + \log n)$ time.

Of course we are interested in a combination of the effects which means multilevel methods on adaptively refined meshes, parallel multilevel methods, parallel adaptive mesh refinement, and finally the combination of all three, see Figure 1.1. We will discuss these aspects in detail in the following chapters.

Finite-Element, Finite-Volume and Finite-Difference methods for the solution of partial differential equations are based on meshes. The solution is represented by degrees of freedom attached to certain locations on the mesh. Numerical algorithms operate on these degrees of freedom during steps like the assembly of a linear equation system or the solution of an equation system. A natural way of porting algorithms to a parallel computer is the data parallel approach. The mesh with attached degrees of freedom is decomposed

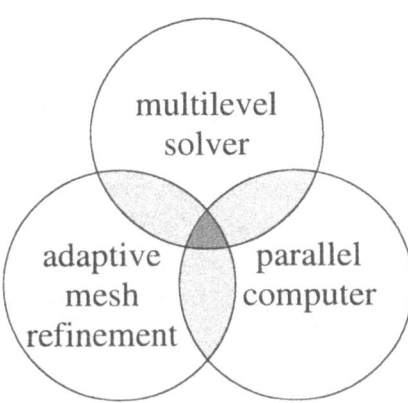

Figure 1.1. *Three ingredients of efficient numerical simulation: adaptive mesh refinement, multigrid equation solver and parallel computing.*

into several partitions and mapped to the processors of the parallel computer. Accordingly the operations on the data are also partitioned. Goals of a partitioning scheme are a balanced load on the processors and little communication between the processors. Sometimes singly-connected partitions are also required. Furthermore, if the partitions are determined during run-time, furthermore a fast partitioning scheme itself is sought. This is for example the case within adaptive mesh refinement of a PDE solver.

A good solution to the partitioning problem is a key point for the efficiency of adaptive parallel codes. The computational load has to be equidistributed over the processors. Here, data has to be transferred between processors during computation and during the mapping of the partitions to processors. Of course, the volume of this data should be low. Furthermore the load balancing and mapping process should be cheaper than the actual computation. Often the last demand is violated, since the load balancing step is applied less often. Consequently some load imbalance and a harder mapping problem results, since more load has to be distributed in the rarer mapping steps.

In this book we propose a space-filling curve enumeration scheme for the parallel load balancing problem. It is based on the principles of self-similarity and scaling invariance. It provides even multilevel locality, i.e. as much locality on each scale as possible. We introduce the space-filling curve schemes and prove some of the properties of the partitions. The scheme is cheap, deterministic, incremental, can be parallelised and provides acceptable partitions. However, even more striking, it seems to be one of the few partitioning methods where quasi-optimal estimates can be shown. We are able to derive sharp estimates both on the partition and on the multilevel algorithms on the partition, which is more than is known about competing graph partitioning load balancing methods so far.

Furthermore, we give a survey of the three main aspects of the efficient treatment of PDEs, that is, discretisation, multilevel solution and parallelisation. We will treat the main features of each of the three distinct topics and cover the historical background and modern developments. We demonstrate how all three ingredients can be put together to give an adaptive and parallel multilevel approach for the solution of PDEs. Error estimators and adaptive mesh refinement techniques for ordinary and for sparse grid discretisations are presented. Different types of additive and multiplicative multilevel solvers are discussed with respect to parallel implementation and application to adaptive refined meshes. Efficiency issues are treated both for the sequential multilevel methods and for the parallel version by hash table storage techniques. Furthermore, space-filling curve enumeration for parallel load balancing and processor cache efficiency are discussed. We will apply the method to elliptic boundary value problems.

We are able to derive estimates for the quality of the partitions by space-filling curves and the load balancing of the numerical algorithms on the meshes. Even for adaptive mesh refinement within certain ranges we are able to prove that the partitions are quasi-optimal, i.e. the cut sizes of the dual graph are only a constant factor away from optimum independent of the mesh size. Hence we obtain asymptotic optimality of the parallel algorithms. This seems to be remarkable in comparison to graph based heuristics, where little is known about the quality.

Furthermore we were able to demonstrate the performance of the method on a range of the world's largest parallel computers, namely ASCI Blue Pacific and a prototype Cray T3E. We complement this data by simulations run on Parnass2, which was the first non-US self-made cluster in the list of the world's largest 500 computers (TOP500).

In chapter 2 we discuss the solution of the equation systems, which arise from the discretisation of partial differential equations. Starting with direct Gaussian elimination and basic iterative methods, we turn to multigrid and domain decomposition methods. Furthermore aspects of linear algebra for sparse grid discretisations are discussed. We introduce the analysis of multigrid and domain decomposition methods within the unifying framework of subspace correction methods due to Bramble [57], Griebel and Oswald [146] and J. Xu [311]. Based on an abstract splitting of function spaces, iterative schemes like additive and multiplicative Schwarz iterations can be defined. The convergence rate of the methods depends on the spectrum of the splittings. Multigrid and multilevel methods can be defined by a nested sequence of spaces, whereas spaces defined on a geometric partition of the domain lead to domain decomposition methods. They allow for an efficient solution of large equation systems obtained by the discretisation of PDEs in $\mathcal{O}(n)$ or $\mathcal{O}(n \log n)$ operations for n equations, which follows from estimates on the abstract subspace splittings. The main motivation for the development of domain decomposition is the need for efficient iterative solvers on parallel computers so that the computational domain is partitioned and mapped onto several processors. However, multigrid methods can also run efficiently in parallel. We conclude the chapter with a discussion of sparse grid discretisations and the solution of equation systems derived therefrom. Just like other higher order discretisations, some of the functions in the Galerkin method do not have a local support. This leads to larger fill-in and to alternative methods both to discretise and to solve the equation systems.

In chapter 3 we discuss the construction of adaptively refined meshes with a posteriori error estimation and adaptive mesh refinement, the discretisation of partial differential equations on adaptively refined meshes and the representation of the meshes. Since a multigrid $\mathcal{O}(n)$ computational complexity for n degrees of freedom is optimal and cannot be beaten by any algorithm with n words of output, there are only two ways for a substantial improvement of a PDE solver. A parallel computer might be able to reduce the sequential complexity with a large number of processors, and an adaptive mesh refinement procedure might reduce the actual number of degrees of freedom n. We will discuss the latter in this chapter, while the parallelisation will be treated in the remaining chapters. The chapter covers first the discretisation of partial differential equations on adaptively refined meshes, where special care is needed to maintain the approximation order for finite differences and the symmetry for finite elements in the presence of hanging nodes. Next standard a posteriori error estimators are reviewed. Different types of meshes require different mesh refinement and mesh manipulation algorithms, be it triangles with bisection or red-green refinement, quad-tree meshes with hanging nodes, or some hybrid meshes. Finally we discuss the representation of adaptive meshes on a computer, both as data structures and algorithmic components accessing them. A structure which seems to be new in the field of partial differential equations is key based addressing. Instead of enumeration or direct references to memory, the abstract interface of a key serves as a clear separation between mesh data and numerical and discrete algorithms. Furthermore key based addressing also simplifies the parallelisation of a code, because keys are indeed portable, while memory references and local indexes are not on a distributed memory computer, as we will see later on. Moreover, we discuss an efficient implementation of key based addressing schemes by means of hashing with a constant time access complexity in the statistical average.

The parallel version of a code with adaptive mesh refinement poses the problem of load balancing. Since it is not known in advance what the refined mesh will look like, an a priori partition and mapping of the mesh to the processors cannot be used. Furthermore the partition of the initial coarse mesh and a good partition of the final adaptive mesh will differ substantially, so that a re-partitioning or dynamic load balancing is necessary at run-time. For this purpose, space-filling curves are introduced in chapter 4. Together with the estimates of this chapter, they are situated right in the centre of this book and are important ingredients for the remaining chapters. We start with a review of classical space-filling curves from its beginnings with the constructions of Peano and Hilbert and examine some of the basic properties including Hölder continuity. Then we discuss the mesh partitioning problem in general. Space-filling curves can be used to enumerate nodes or elements of a mesh, since they cycle through the whole domain. This enumeration can be used to sort the elements and to assemble them to connected domains. Such domains serve as mesh partitions. We will prove that the partitions are indeed well suited for parallel computing in the sense that the cut sizes of the dual graph are of quasi-optimal size. In other words, the surface of the partitions can be bounded in terms of the volume of the partitions with bounds just a constant factor larger than the optimum, independent of the mesh size. Initially we will show this estimate just for uniformly refined meshes, where it is based on the Hölder continuity of the function with exponents $1/d$. Later on, we develop a set of refinements of the proof to cover also adaptively refined meshes. For this purpose we introduce criteria for adaptively refined meshes which we call β and γ refinement. The results show that usual adaptive refinement as well as best n-term approximation for solutions with point-singularities allow for indeed quasi-optimal space-filling curve partitions. Furthermore, the refinement toward higher dimensional manifolds like edge singularities may also lead to quasi-optimal partitions, dependent on the dimension d of the domain and in the case of best n-term approximation also on some parameters of the underlying Besov spaces. However, we also show that refinement toward manifolds with co-dimension one or adaptively refined meshes with arithmetic progression of nodes cannot lead to optimal estimates. This does not mean that all partitions of such meshes exhibit bad cut-sizes, but some partitions do, as can be shown also experimentally. In order to construct good partitions also in these cases, we propose the construction of anisotropic space-filling curves, which are aligned to the refinement manifold. The quasi-optimal estimates also carry over to the analysis of algorithms on the partitions like iterative solvers or a complete multigrid cycle, so that asymptotic parallel efficiency can be shown. We conclude the chapter on space-filling curves by a discussion of sparse grids. Here even optimal partitions of sparse grids do exhibit larger cut sizes than standard discretisations due to the global support of some of the Galerkin function or the global size of the finite difference stencils.

In chapter 5 we combine multilevel methods, adaptive mesh refinement and space-filling curve partitioning of the previous chapters. First of all we discuss multigrid and multilevel methods on adaptively refined meshes. However, the efficient $\mathcal{O}(n)$ implementation of multilevel methods on a sequence of adaptively refined meshes requires more care, since the number of nodes per mesh need not increase geometrically. Alternative multilevel methods to cope with the problem are for patch-wise adaptive discretisations by Brandt [64], local multigrid where smoothers are restricted to a neighbourhood of nodes of a level by Rivara [255], hierarchical basis methods by Bank, Dupont and Yserentant [20], and additive multilevel methods. Moreover, we discuss the combinations of parallelism and multigrid methods. While standard multigrid can be parallelised by the combination of parallel smoothers on each level and parallel mesh transfer operators together with a parallel or sequential coarse mesh solver, again alternative multigrid methods like the BPX preconditioner by Bramble, Pasciak and Xu [63] exist. Finally we discuss the combination of adaptive mesh refinement and parallelisation. This topic includes the mesh partition problem, re-partitioning and dynamic load balancing on a parallel computer. Unfortunately, solutions available for shared memory processors like by Leinen [197] do not carry over to distributed memory environments. We also cover parallel multigrid methods on the parallel adaptive mesh refinement. For this purpose we again propose key based addressing, with processor independent keys and a parallel implementation of the underlying hash storage scheme. The parallel hashing can be tightly integrated into the deterministic space-filling curve partitioning scheme. In this way we are able to simplify the implementation of an adaptive parallel multigrid method, compared to previous efforts by De Keyser and Roose [100], Bastian [26], Stals [285], and Mitchell [214]. Furthermore, the highly efficient and parallel load balancing can be used effectively to balance every cycle of the procedure, which so far has been possible only for shared memory computers.

In the final chapter 6 on numerical experiments we demonstrate how space-filling curve partitions can be used within an adaptive multilevel method and a sparse grid PDE solver. We cover a range of applications, from the elliptic Poisson equation and linear elasticity to the convection-diffusion equation and a high dimensional problem, where sparse grids are needed. The experiments are carried out on several parallel computer platforms, namely the Parnass2 cluster with up to 144 processors which we built at our department [272], ASCI Blue Pacific at Lawrence Livermore National Laboratories with up to 1280 processors and several Cray T3E computers with up to 1400 processors. We are able to compare the same code on several platforms and for a large range of processors, which demonstrates the scalability of the approach. We cover two and three dimensional problems, uniformly and adaptively refined meshes and demonstrate that the asymptotic estimates on graph cut sizes of space-filling curves are indeed useful also for realistic test cases. Moreover, we see how three dimensional problems are harder to parallelise than two dimensional problems, how sparse grid algorithms are harder to parallelise than standard discretisations and how adaptively refined meshes are harder to parallelise than uniformly refined meshes. We also observe how algorithms and parallel computers behave in the region of thousands of processors and how this contrasts to observations on smaller scales. We will conclude that the concept of space-filling curves does indeed work well, simplifies implementation details and for a first time allows for real dynamic load balancing of adaptive computations.

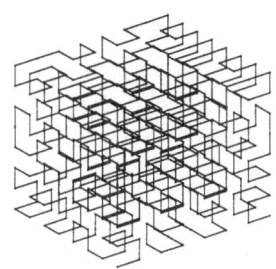

Chapter 2

Multilevel Iterative Solvers

The numerically most expensive part of the solution of boundary value problems is usually the solution of linear equation systems. The effort needed for mesh generation varies substantially between almost negligible for simple geometric domains to months of manual work for good meshes of highly complex shape domains. Adaptive mesh refinement and the assembly of equation systems can be performed with low complexity on a computer. However, the major part of computing time is spent on the solution of the equation systems. Hence it is important to use fast methods for this task.

We will discuss several classes of methods, ranging from direct solvers to multigrid and domain decomposition schemes. The computational complexity of the different methods varies dramatically from $\mathcal{O}(n)$ to $\mathcal{O}(n^3)$ for the solution of n equations, but we will also discuss the applicability of the solvers depending on the boundary value problem and the problem size. Of course we emphasise multilevel methods, which are the methods of choice with asymptotically linear complexity. We present the abstract framework of subspace correction algorithms with its theory, and we describe multigrid methods as well as overlapping and non-overlapping domain decomposition methods.

The chapter is structured as follows. We begin with a review of classical and newer developments of direct and iterative solvers. From dense matrices and banded matrices to sparse matrices the range of problem sizes increases for direct solvers. However, especially for three-dimensional problems the matrix fill-in is prohibitively large, so that direct methods are not applicable for large scale problems due to memory consumption and computational complexity. Alternative iterative solvers do not suffer from the memory problem, but still can have high complexity. However, more advanced schemes like modified ILU or the Fast Fourier Transform (FFT) do have a competitive complexity. The main drawback is their limited applicability, so that domains other than the unit square and varying coefficients cannot be solved with the same complexity.

Hence we discuss multigrid and domain decomposition schemes, which are more efficient schemes than many of the iterative methods and have a wider applicability than methods tuned for the Poisson equation on the unit square. We introduce the methods in the general framework of subspace correction schemes, where a convergence theory of abstract Schwarz iterations is developed. Based on the spectrum of the variational form, convergence rates of the additive and multiplicative Schwarz iteration are given. Within this general framework both classical multigrid and multilevel methods and various domain decomposition methods can be analysed.

We give a short review of the (historical) development of multigrid methods for elliptic problems and show how they fit into the framework. Here it is interesting to compile the different techniques developed so far, for example to tackle the problem of robustness of the solvers with respect to the coefficient functions of the differential operator. Domain decomposition methods can be categorised into overlapping Schwarz methods, which fit directly into the subspace correction framework, and non-overlapping Schur complement based methods. Here the analysis of domain decomposition preconditioners for the Schur complement again is done by the subspace correction framework.

The chapter is concluded by a section on sparse grid discretisations and solvers. Sparse grids can be considered as a specific way to discretise partial differential equations. Based on a one-dimensional multi-resolution analysis or a (pre-) wavelet basis, a multidimensional approximation for functions with bounded mixed derivatives is set up. The purpose of this discretisation is to approximate the solution of a partial differential equation by fewer degrees of freedom than by a classical mesh based approximation. However, many of the algorithms discussed so far for the solution of the linear equation systems are not applicable due to the global interaction of basis functions and the sparsity pattern of the stiffness matrix. Hence, iterative solvers tailored directly for the sparse grid discretisations are discussed. Again, the type of solver depends on the discretisation, so that the combination technique, the Galerkin and the finite difference discretisations on sparse grids all require different solvers.

2.1 Direct and Iterative Solvers

In this section we will introduce direct and standard iterative methods for the solution of linear equation systems. The main purpose of this review is to demonstrate the limits of current methods, both in the size of equations systems and in the range of problems which can be solved. While direct solvers for dense matrices are immediately limited by the available memory even for input data, banded and current sparse matrix solvers are able to push the limits further. However, three-dimensional problems lead to more fill-in, so that large problems can only be solved by iterative methods. In addition to the well known complexity estimates, we also give a comparison of estimated actual execution times on sequential and parallel machines, taking the different levels of computational efficiency of the classes of algorithms into account. It turns out that advanced iterative methods like Krylov iterations preconditioned by a modified ILU scheme or by a Fast Fourier Transform (FFT) are highly competitive as long as they can be applied to the problem in question. However, if we consider a computational domain different from the unit square or a differential operator different from the Laplacian, both methods have to be modified and become less efficient.

2.1.1 Dense Linear Algebra

The straightforward methods for the solution of linear equation systems are direct dense matrix linear algebra methods like Gaussian elimination in general, Cholesky factorisation for symmetric matrices and QR-decomposition as an extremely stable decomposition method. It is well known that the computational complexity of $\mathcal{O}(n^3)$ is prohibitively high for large problems. Nevertheless, the methods are among the most efficiently implemented numerical algorithms. Code optimisation over several decades led to sequential and parallel codes close to the peak performance of a broad range of computers.

Direct solvers with pivot search for numerical stability require $\mathcal{O}(n^2)$ storage, which can be the matrix storage. The operation count for Gaussian elimination is $\frac{2}{3}n^3 - n^2$ plus lower order terms. On a hypothetical sequential computer with 1 GFlop/s peak performance and 1 Gbyte main memory, the largest equation system that fits into memory is roughly of size $n = 11,500$ in double precision (8 byte) arithmetic. If we assume an efficiency of 90% peak performance, the solution requires 19 minutes. A smaller system of $n = 1000$ could be solved in 0.74 seconds. A 1024 processor parallel computer built of this hypothetical sequential computer would be able to solve equation systems up to $n = 370,000$ which would take more than 10 hours. This seems to be realistic, see also van de Geijn [296], Anderson et al. [5], and Greer and Henry [132]. Since this parallel computer would be a 1 TFlop/s machine, we see immediately that dense direct solvers are severely limited in n. Currently there is no chance to approach a number of equations of say $n = 10^6$. Nevertheless, dense linear algebra routines serve as highly efficient computational kernels within larger implementations.

A more economical way of storing matrices is a band storage scheme. The Gaussian elimination without pivot search gives an LU factorisation with banded upper and lower triangular factors. The implementation of banded algorithms is close to the full matrix implementation, but uses less memory and fewer operations. For a symmetric bandwidth of m, i.e. $a_{i,j} = 0$ for $|i-j| > m$ the matrix requires $(2m+1)n - m(m+1)$ values storage. A Gaussian elimination accounts for $2nm^2 + 11nm$ operations plus lower order terms. The main question is, how large is the bandwidth of a matrix? This depends highly on the problem and its discretisation. Let us consider a regular mesh, Laplace operator and linear finite elements or small difference stencils. The discretisation of the two-dimensional unit square gives an approximate number of equations $n = h^{-2}$ and a minimum bandwidth of $m = \sqrt{n}$ for lexicographical ordering. We get $\mathcal{O}(2n^{3/2})$ storage and $\mathcal{O}(2n^2)$ number of operation cost, which is a substantial improvement compared to the dense matrix case. However, higher dimensional cases do not look that good: in general on the d dimensional unit cube we get a bandwidth of $m = n^{(d-1)/d}$. Hence the storage required is of the order $\mathcal{O}(2n^{1+(d-1)/d})$ and $\mathcal{O}(2n^{1+2(d-1)/d})$ number of operations are required for the solution of the equation system.

The hypothetical sequential computer can now solve problems of roughly up to size $n = 50,000$ stemming from the discretisation of the Laplacian on the unit cube, in $3\frac{1}{2}$ minutes. The hypothetical parallel computer could solve problems up to $n = 3.1 \cdot 10^6$ for $d = 3$ in roughly an hour. We see that the limit of problem sizes that are computable is shifted by the step from dense to banded matrices by some factor, while at the same time the computations at the limit size are much faster, compare also Figure 2.1 and experiments in Bruaset et al. [74].

2.1.2 Sparse Matrix Solver

The implementation of direct Gaussian elimination can be further improved for the case of finite elements or finite differences. Since the performance of a factorisation algorithm crucially depends on the size of the factors, i.e. the number of matrix entries plus the matrix fill-in, one has to reduce the size. Node numbering strategies can order the unknowns so that the number of non-zeros in the matrices L and U is minimised. However, a band-matrix structure is not well suited to store the matrices in this order, because it stores values which are known to be zero in addition. A general sparse matrix scheme is more appropriate. The solution process consists of three stages,

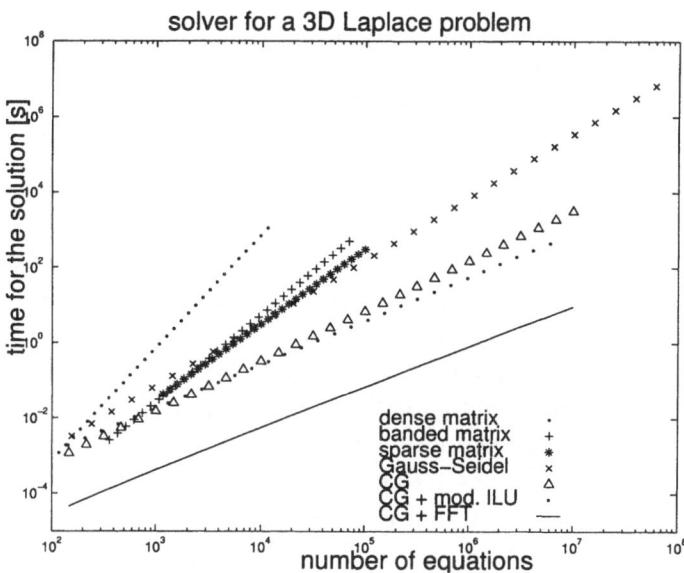

Figure 2.1. *Performance of linear equation solvers for a three-dimensional model problem on a model computer.*

		$d = 2$	$d = 3$
Gaussian elimination	dense matrix	$\mathcal{O}(\frac{2}{3}n^3)$	$\mathcal{O}(\frac{2}{3}n^3)$
	banded matrix	$\mathcal{O}(n^2)$	$\mathcal{O}(n^{7/3})$
	sparse matrix	$\mathcal{O}(n^{3/2})$	$\mathcal{O}(n^2)$
iterative methods	Jacobi, Gauss-Seidel, ILU	$\mathcal{O}(n^2)$	$\mathcal{O}(n^{5/3})$
	SOR, modified ILU, CG	$\mathcal{O}(n^{3/2})$	$\mathcal{O}(n^{4/3})$
preconditioner	modified ILU	$\mathcal{O}(n^{5/4})$	$\mathcal{O}(n^{7/6})$
	FFT	$\mathcal{O}(n \log n)$	$\mathcal{O}(n \log n)$

Table 2.1. *Complexity of solvers for two and three dimensional model problems.*

namely re-ordering, symbolic computation of the matrix patterns and the fill-in, and the numerical factorisation itself. If additional pivot search is needed for numerical stability, the matrix permutation usually increases the fill-in and the re-ordering becomes less effective. Hence there are recent trends to limit pivoting for performance reasons, especially in parallel sparse matrix solvers. Numerical stability of a factorisation can be improved later on by iterative defect correction of the solution. This can be interpreted also as an iterative solver with Gaussian elimination as a preconditioner.

For obvious reasons it is more difficult to discuss the complexity and efficiency of sparse matrix solvers than of dense matrix solvers. One observes large variations of the performance of codes applied to standard benchmark sets of matrices. However, for some model problems of the solution of boundary value problems, data is available. In the case of uniform meshes for the unit square or unit cube, nested dissection ordering is known to be optimal, see Figure 2.2. For the discretisation of the Laplacian on the unit cube, the number of operations is approximately $12n^2$ compared to $\mathcal{O}(n^{2\frac{1}{3}})$ for band matrices. Hence for $d \geq 2$, sparse matrices pay off for the model problem. In general, sparse matrices are much more flexible and cover matrices whose band matrix storage is not very economical. Due to the management of the sparsity pattern in addition to the values of the matrix elements, the solvers are slightly less efficient than the dense matrix implementations both in storage and in memory. For a comparison on our hypothetical computer we extrapolate some performance data from the parallel sparse matrix package 'Spooles' by Ashcraft and Grimes [8], 'Mumps' by Amestoy, Duff and L'Excellent [4], and 'SparseLU' by X. Li and Demmel [200]. If we assume a sequential efficiency of 40%, the example of $n = 50,000$ would require 17 seconds solution time. However, it is about to fill the 1 Gbyte of main memory due to the sparse matrix storage of the factorisation. Typical results for the parallel execution of sparse matrix solvers suggest that the parallel efficiency for the 1024 processor machine would be poor, maybe in the range of $1 - 2\%$. This seems to be true both for one-dimensional matrix partitions onto the parallel processors, which support pivoting but are known to be sub-optimal, see Schreiber [268], and for two-dimensional matrix partitions. The main memory of the parallel hypothetical computer would be sufficient for $n = 5 \cdot 10^6$ at an execution time of four hours.

The comparison with banded matrix direct solvers, Figure 2.1, shows that the difference is not that large for the model problem. This is no longer true for discretisations on general, unstructured meshes, where the band width m can be prohibitively large. Furthermore there are many different sequential, shared memory parallel and message-passing parallel sparse matrix solver implementations which differ widely in performance. Some overview can be found in George and Liu [125, 126]. Parallel software packages include the Spooles, Mumps, SparseLU, 'Oblio' by Dobrian, Kumfert and Pothen [103], 'Mcsparse' by Gallivan, Marsolf and Wijshoff [122], codes by Ashcraft, Eisenstat and Liu [7] and by Fu, Jiao and Tang [120] and some others. The codes differ in the communication and factorisation mode such as multi-frontal and fan-in, in the matrix formats, in the symbolic factorisation and in the pivoting. Note that

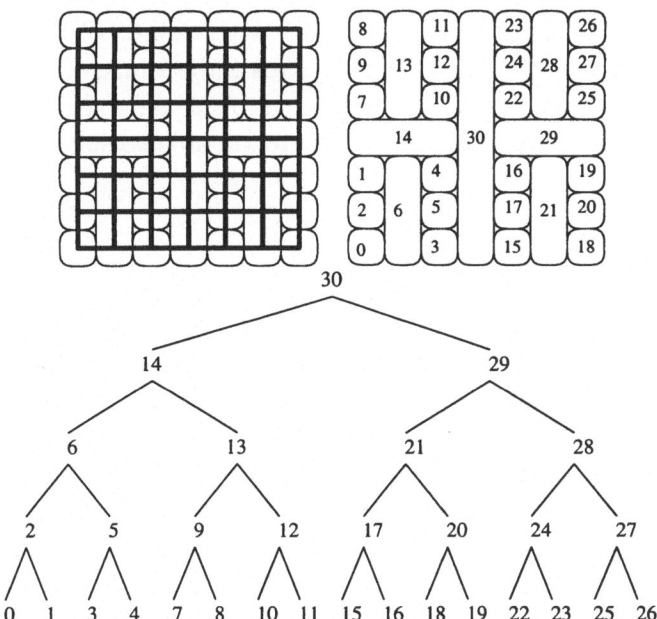

Figure 2.2. *Nested dissection node numbering (right) and elimination tree (bottom) for a two-dimensional uniform square mesh (left).*

the nested dissection ordering of the two-dimensional finite difference discretisation of the Laplace operator on the unit square gives a $\mathcal{O}(n \log n)$ complexity as a special case (Buneman, total reduction [77]).

An algorithm specialised for uniform meshes of the d-dimensional unit cube is based on the Fourier transform. The efficient implementation of the FFT in $\mathcal{O}(n \log n)$ gives a decomposition into the eigenmodes of the Laplacian on the cube. The solution can be found by a real Fourier transform, scaling the modes and a back transform. The procedure can also be used as a preconditioner in case the boundary value problem differs from the model problem, e.g. due to mild variations of the coefficients or boundary conditions. For a comparison the exact performance numbers of an FFT implementation are of interest, which depend on the factorisation of n into primes. However, the original base two Cooley-Tukey FFT has a work count of $5n \log_2 n + \mathcal{O}(n)$. On our hypothetical computer, we can expect an efficiency of an implementation like 'FFTW' by Frigo and Johnson [119] of about 40%–70% for small n and 20%–40% for large n. The main reason for this difference is the speed of processor caches versus main memory. Large problems use the whole main memory, while relatively few operations are performed. Using the full main memory, we might be able to store and solve a problem with $n = 60 \cdot 10^9$ in one day, while a smaller problem of $n = 50,000$ only takes only 16ms. Of course this does not include any matrix storage or reference. A parallel FFT implementation on our hypothetical parallel computer might be able to solve a problem of size $n = 60 \cdot 10^{12}$ in 30 hours.

2.1.3 Iterative Solvers

The application of direct solvers is limited for large problems by memory constraints. In addition to the matrix, the factorisation requires all available memory. Even out-of-core methods which store the factors to disk are limited by computing time. The alternative FFT solver is limited to cube shaped Laplace-type problems. Hence a general memory efficient solver is needed. Iterative methods provide such a class of methods. Besides the matrix, only a few vectors are needed. Most of the methods can be implemented in a matrix-free style, i.e. no explicit matrix storage is needed. This can be advantageous, if the matrix vector product Ax can be implemented without additional memory as for many finite difference stencil discretisations.

We will briefly discuss several iterative schemes. The basic iterative methods are the Jacobi and Gauss-Seidel iterations, based on updates of type $x_i^{\text{new}} = (f_i - \sum_j A_{i,j} x_j^{\text{old}})/a_{i,j}$, with a single update for all x or successive updates for each i. An iterative method can reduce the error successively, but does not give the solution after a finite number of operations. Of course machine rounding will lead to a fix-point, but a useful solution is obtained much quicker. We can either prescribe an error tolerance or some residual tolerance, which implies a number of iteration steps. The tolerance can be chosen as a fixed number, i.e. related to the machine precision. However, it might be even more efficient to relate the error to the discretisation error, since an iteration error sufficiently smaller than the discretisation error would not be visible in the numerical solution. A larger number of unknowns means a finer mesh, a smaller discretisation error and a larger number of iterations. The complexity depends on whether we consider a fixed or a discretisation dependent tolerance.

We consider the test problem, the Laplacian on the unit square and the unit cube. The convergence rate of the Jacobi and Gauss-Seidel iterations depend on the spectrum of the operator, which itself depends on the mesh size. The convergence rate of the Jacobi iteration is

$$\rho = \cos \pi h = 1 - \frac{\pi^2}{2} h^2 + \mathcal{O}(h^4)$$

with mesh size h and of the Gauss-Seidel iteration slightly better as $\rho =$

$1 - \pi^2 h^2 + \mathcal{O}(h^4)$. We consider finite element or finite difference discretisations so that a single iteration requires $\mathcal{O}(n)$ operations and $\mathcal{O}(n)$ storage. A dense matrix would for example require both more operations per iteration and more storage. The mesh size h of the model problem is of the order $h = n^{-1/d}$, hence the number of iterations needed to achieve a fixed tolerance ϵ is

$$m = \mathcal{O}(h^{-2}|\log \epsilon|) = \mathcal{O}(n^{2/d}|log\epsilon|) \,.$$

The work count for a fixed accuracy is $mn = \mathcal{O}(n^{1+2/d}|\log \epsilon|)$, while a solution up to discretisation error with $\epsilon = h^2$ gives a work count of $\mathcal{O}(n^{1+2/d} \log n)$ all with $\mathcal{O}(n)$ storage.

The Gauss-Seidel iteration can be improved by over-relaxation. This method is called SOR and has a convergence rate for the model problem of

$$\rho = \frac{1 - \sin \pi h}{1 + \sin \pi h} = 1 - 2\pi h + \mathcal{O}(h^2)$$

which reduces the number of iterations to

$$m = \mathcal{O}(h^{-1}|\log \epsilon|) = \mathcal{O}(n^{1/d}|\log \epsilon|)$$

and the work count to $\mathcal{O}(n^{1+1/d}|\log \epsilon|)$ with fixed accuracy and $\mathcal{O}(n^{1+1/d} \log n)$ for the solution up to discretisation error.

The theory for Krylov iterations differs substantially from the analysis of the basic iteration schemes so far. However, the version for symmetric positive definite matrices, the conjugate gradient (CG) method, has a convergence rate similar to the SOR iteration for the model problem. The error decays roughly as $\rho \approx 1 - 2/\sqrt{\kappa}$ with the condition number of the matrix κ. The condition for a second order differential operator is of the order $\kappa = \mathcal{O}(h^{-2})$. Hence the number of iteration is

$$m = \mathcal{O}(\sqrt{\kappa}|\log \epsilon|) = \mathcal{O}(h^{-1}|\log \epsilon|) = \mathcal{O}(n^{1/d}|\log \epsilon|)$$

and the work count has the same order as that of the SOR iteration. The conjugate gradient method requires slightly more storage for five auxiliary vectors. It does not work for non-symmetric matrices, where Krylov methods like GM-Res, BiCG or CGS come into play. The exact behaviour of the conjugate gradient is determined by the spectrum of the matrix so that convergence may be faster for multiple or clustered eigenvalues.

The main reason to use the conjugate gradient method for the solution of large systems of equations is the possibility to combine it with some preconditioner. Given some self-adjoint operator B, e.g. spectrally equivalent to the

inverse A^{-1} of the matrix A, the convergence of the conjugate gradient method
preconditioned by B is determined by the condition number $\kappa(B^{-1}A)$.

The iterative methods so far, Jacobi and Gauss-Seidel, can be used as a
preconditioner for the conjugate gradient method. However, the complexity
of the conjugate gradient method of the solver is not improved for the model
problem, which does not mean that such preconditioners are not useful for
other boundary value problems. Another class of iterative solvers that can be
used more successfully is the incomplete LU-factorisations (ILU). Instead of
a full Gaussian elimination, the factors are restricted to a prescribed sparsity
pattern, often related to the sparsity pattern of the matrix itself. In this way
the factorisation is not exact and the method can be used as an iterative
method. The amount of work and the storage for standard decompositions
and our model problem is $\mathcal{O}(n)$ for the factorisation and a single iteration.
The convergence rate of the iteration is of the order $\rho = 1 - ch^2$. There
is a modification of the ILU scheme by Gustafsson [158] which improves the
convergence rate for the model problem to $\rho = 1 - ch$. The idea is to equilibrate
the matrix rows of $R := LU - A$ to sum zero, i.e. set the diagonal diagonal
entry $r_{i,i} = -\sum_{j \neq i} r_{i,j}$. This scheme can be used as a preconditioner of the
conjugate gradient method with a convergence rate of $\rho = 1 - c\sqrt{h}$. The
number of iterations is consequently

$$m = \mathcal{O}(h^{-\frac{1}{2}}|\log \epsilon|) = \mathcal{O}(n^{\frac{1}{2d}}|\log \epsilon|)\,,$$

which leads to a work count of $\mathcal{O}(n^{1+\frac{1}{2d}}|\log \epsilon|)$ or $\mathcal{O}(n^{1+\frac{1}{2d}})$ depending on the
termination criterion. With the exception of the specialised FFT, this is the
fastest classical linear equation solver. However, the modified ILU version
is effective only for certain classes of problems. As soon as we change the
operator or only the type of the discretisation, the superior convergence rate
is lost. In general ILU solver or preconditioner are always used, if robustness
of the solver is sought. The decomposition is applicable to a wide variety of
boundary value problems and discretisations and often gives good results.

2.2   Subspace Correction Schemes

As we saw, both direct and standard iterative solvers are limited in problem
size and applicability. More advanced iterative solvers and preconditioners are
based on domain decomposition methods and multilevel methods. Now the
differential operator and the discretisation come into play. The domain Ω can
be discretised at different mesh scales h which leads to multilevel methods,
or it can be decomposed into overlapping or disjoint partitions which leads

to domain decomposition methods. The general analysis of the methods is based on the decomposition of the function space into several spaces of smaller dimension, which we will discuss in the next section. Generally we obtain constant or poly-logarithmic convergence rates ρ with a work count of $\mathcal{O}(n)$, which gives asymptotically optimal linear equation solvers. The methods can be interpreted and analysed as subspace correction schemes.

The convergence of multigrid methods can be analysed by classical multigrid theory by Hackbusch [163, 165], Wesseling [304], Braess [52], and Bank and Dupont [19] or by the additive theory by J. Xu [310], Bramble, Pasciak and Xu [63], Bramble, Pasciak, Wang and Xu [62], Oswald [228], Zhang [318], and Dahmen and Kunoth [98]. The latter leads to a regularity free theory, which can be generalised to a theory of subspace correction schemes. The analysis of abstract additive and multiplicative Schwarz methods also includes various domain decomposition methods.

Subspace correction methods are techniques for the numerical solution of equation systems which arise in the discretisation of partial differential equations. Here, the function space in which the solution is sought is decomposed into several smaller spaces. The framework of subspace correction schemes contains both classes of optimal and almost optimal order iterative solvers, multilevel and domain decomposition methods, which will be discussed in more detail in the subsequent sections.

Let V be some fixed (already finite-dimensional) Hilbert space. The scalar product in V is denoted by $(\cdot, \cdot)$. We consider a linear elliptic self-adjoint partial differential equation, which can be written after discretisation with respect to V in operator notation as $Au = f$ with $A : V \to V$ denoting the corresponding symmetric positive definite (s.p.d.) operator acting on V. The associated standard weak formulation is: *find $u \in V$ such that*

$$a(u, v) = f(v) \qquad \forall\, v \in V \tag{2.1}$$

with positive definite symmetric bilinear form $a(u, v) = (Au, v)$, $u, v \in V$ and linear functionals $f(v) \equiv (f, v)$.

2.2.1 Subspace splitting and abstract algorithms

Now, consider an arbitrary additive representation of V by the sum of a finite number of subspaces $V_j \subset V$, which are not necessarily disjoint:

$$V = \sum_{j=0}^{J} V_j \,. \tag{2.2}$$

More precisely, this means that any $u \in V$ possesses at least one representation $u = \sum_{j=0}^{J} u_j$ where $u_j \in V_j$ for all $j = 0, \ldots, J$. Suppose that the V_j are equipped with auxiliary s.p.d. forms $b_j(u_j, v_j) = (B_j u_j, v_j)$ given by the s.p.d. operators $B_j : V_j \rightarrow V_j$. These forms might model approximative solvers used on the subspaces, i.e. B_j^{-1} is an approximative inverse for A_j, the restriction of A to V_j.

We can rewrite problem (2.1) as:

find $u \in V$ such that

$$Pu = \phi \qquad (P \equiv \sum_{j=0}^{J} T_j : V \rightarrow V) \tag{2.3}$$

with the operators $T_j : V \rightarrow V_j$ given by the variational problems

$$b_j(T_j u, v_j) = a(u, v_j) \qquad \forall v_j \in V_j \tag{2.4}$$

and $\phi = \sum_{j=0}^{J} \phi_j$ with $\phi_j \in V_j$ defined via the projection

$$b_j(\phi_j, v_j) = f(v_j) \qquad \forall v_j \in V_j , \tag{2.5}$$

which is analogous to the auxillary problem Equation (2.4).

In operator notation $T_j = B_j^{-1} Q_j A$ where $Q_j : V \rightarrow V_j$ denotes the ortho-projection onto V_j with respect to the scalar product $(\cdot, \cdot)$. Since

$$P = \sum_{j=0}^{J} T_j = \left(\sum_{j=0}^{J} B_j^{-1} Q_j \right) A \equiv C \cdot A , \tag{2.6}$$

the switching to formulation of Equation (2.3) can be viewed as preconditioning strategy with the preconditioner C for the original problem (2.1). Note that the formulation (2.3) is nothing but the additive Schwarz formulation of problem (2.1), which has already been treated by many authors, see Dryja [110, 111], Dryja, Smith and Widlund [109], J. Xu [311], Yserentant [315], and Zhang [318] for further references.

Now we turn to the additive and multiplicative variants of the Schwarz iteration associated with the splitting. The *additive* subspace correction algorithm (J. Xu [311]) associated with the splitting Equation (2.2) and its associated auxiliary forms b_j is defined as the Richardson method applied to Equation (2.3):

$$u^{(l+1)} = u^{(l)} - \omega \cdot (Pu^{(l)} - \phi) = u^{(l)} - \omega \cdot \sum_{j=0}^{J} (T_j u^{(l)} - \phi_j) , \quad l = 0, 1, \ldots . \tag{2.7}$$

Here $u^{(0)} \in V$ is any given initial approximation to the solution u of Equations (2.1) and (2.3), respectively, and ω is a relaxation parameter. It is easy to see that the associated error propagation operator is $I - \omega \sum_j T_j$. In practice, the additive method serves as a preconditioner and the Richardson method is often replaced by a conjugate gradient iteration.

In contrast to the parallel incorporation of the subspace corrections $r_j^{(l)} = T_j u^{(l)} - \phi_j$ into the iteration (2.7), the *multiplicative* algorithm uses them in a sequential way:

$$v^{(l+(j+1)/(J+1))} = v^{(l+j/(J+1))} - \omega \cdot (T_j v^{(l+j/(J+1))} - \phi_j), \qquad (2.8)$$

with $j = 0, \ldots, J$, $l = 0, 1, \ldots$. Here the corresponding error propagation operator looks like $\prod_j (I - \omega T_j)$. This explains why the algorithm is called *multiplicative*.

Following Griebel [137, 138], it is interesting to rewrite the iterations (2.7) and (2.8) as follows: we introduce the Hilbert space

$$\tilde{V} = \{\tilde{u} = \{u_j\} : \sum_j b_j(u_j, u_j) \leq \infty\}, \qquad (\tilde{u}, \tilde{v})_{\tilde{v}} = \sum_j b_j(u_j, v_j)$$

which is just the usual Cartesian product of the Hilbert spaces V_j, i.e.

$$\tilde{V} = V_0 \times V_1 \times \ldots \times V_J.$$

We introduce the operator $\tilde{P}$ (in matrix representation) with respect to the *coordinate* spaces V_j and decompose it into lower triangular, diagonal and upper triangular parts

$$\tilde{P} = \{T_{i,j}\}_{i,j=0}^J = \tilde{L} + \tilde{D} + \tilde{L}^*, \qquad T_{i,j} \equiv T_j|_{V_j} : V_j \to V_i \qquad (2.9)$$

where $L_{j,i} = T_{j,i}$ for $j < i$ and $D_{j,i} = T_{i,i}$ while all other entries are zero operators between the respective subspaces. We further introduce the summation operator $S : \tilde{u} \in \tilde{V} \to u \equiv S\tilde{u} = \sum_j u_j$ which is a linear operator from $\tilde{V}$ onto V. Finally, to any linear continuous functional f on V, we associate the elements $\tilde{\phi} \in \tilde{V}$ by defining $\tilde{\phi} = \{\phi_j\}$ with ϕ_j from the auxillary problem Equation (2.5).

Now, we consider the problem: *find $u = S\tilde{u} \in V$ where $\tilde{u} \in \tilde{V}$ is such that*

$$\tilde{P}\tilde{u} = \tilde{\phi}. \qquad (2.10)$$

Here, the solution u is unique, while $\tilde{u}$ is not, since different decompositions are allowed. It can be shown that problem (2.10) has the same solution u as (Equations 2.1) and (2.3), see Griebel and Oswald [146] for further details.

If we now define the Richardson (or Jacobi-type) iteration with respect to problem (2.10)

$$\tilde{u}^{(l+1)} = \tilde{u}^{(l)} - \omega \cdot (\tilde{P}\tilde{u}^{(l)} - \tilde{\phi}) \quad l = 0, 1, \ldots$$

and the SOR-like iteration

$$(\tilde{I} + \omega \cdot \tilde{L})\tilde{u}^{(l+1)} = (\tilde{I} - \omega \cdot (\tilde{D} + \tilde{L}^*))\tilde{u}^{(l)} + \omega \cdot \tilde{\phi}, \quad l = 0, 1, \ldots$$

then they will recover the iterative procedures (2.7) and (2.8) in the sense that they satisfy $u^{(l)} = S\tilde{u}^{(l)}$, whenever this relation has been fulfilled for the starting iterate, i.e. $l = 0$. A consequence of this observation, which was made in Griebel [137, 138] and Griebel and Oswald [144], is that the analysis of the methods (2.7), (2.8) can be carried out in almost the same spirit as in the traditional block-matrix situation if one uses the formulation (2.10). The old proofs are sometimes tricky, including the original one in Bramble, Pasciak, Wang and Xu [61, 62], J. Xu [311], and Bramble and Pasciak [58], see also the comments on this point by Yserentant [315].

2.2.2 Theory

Now, we define a norm $|||.|||$ on V by

$$|||u|||^2 = \inf_{\substack{u_j \in V_j \,:\, u = \sum\limits_{j=0}^{J} u_j}} \sum_{j=0}^{J} b_j(u_j, u_j) , \qquad (2.11)$$

and introduce the positive and finite values

$$\lambda_{\min} = \inf_{u \in V, u \neq 0} \frac{a(u, u)}{|||u|||^2} , \quad \lambda_{\max} = \sup_{u \in V, u \neq 0} \frac{a(u, u)}{|||u|||^2} . \qquad (2.12)$$

The quantity

$$\kappa \equiv \frac{\lambda_{\max}}{\lambda_{\min}} \qquad (2.13)$$

will be called the condition number of the splitting (2.2). It can be seen easily that it is equivalent to the condition number of $C \cdot A$ of the preconditioner Equation (2.6). The spectrum of the operator $P = \sum_{j=0}^{J} T_j$ is given by the constants from Definition (2.12) compare for example Matsokin and Nepomnyaschikh [207], Bjørstad and Mandel [44], Widlund [306], and Zhang [318] and see also the Fictitious Space Lemma by Nepomnyaschikh [219, 220]. Note that the condition number will not change if we change the order of the subspaces.

The problems (2.1) and (2.10) are equivalent and have a unique solution.

Lemma 2.1. Suppose that V is finite-dimensional, and that the splitting (2.2) is finite. Let the characteristic numbers $\lambda_{\max}$, $\lambda_{\min}$, and κ of the splitting be given by *Definitions (2.12), (2.13)*.

(a) *The problems (2.1) and (2.10) have the same (unique) solution.*

(b) *The operator $P = \sum_{j=0}^{J} T_j$ is symmetric positive definite on V w.r.t. $(\cdot, \cdot)_V$ and the minimal interval containing its spectrum is given by the constants from Definition (2.12):*

$$\text{spectrum}(P) \subset [\inf_{u \in V : a(u,u)=1} a(Pu, u), \sup_{u \in V : a(u,u)=1} a(Pu, u)] = [\lambda_{\min}, \lambda_{\max}]$$
(2.14)

Thus, the condition number $\kappa(P) = \|P\|_{V \to V} \cdot \|P^{-1}\|_{V \to V}$ of P coincides with the condition number κ of the splitting (2.2) defined by the Definition (2.13).

This lemma is simple but useful: 2.1(b) provides conditions on the splitting (2.2) and the choice for the forms b_j under which the new problem (2.3) is well-conditioned.

Lemma 2.1(b) has many authors,
Its proof can be given in a few lines using the explicit formula

$$a(P^{-1}u, u) = |||u|||^2 \quad , \qquad u \in V,$$

see, e.g. Widlund [306], J. Xu [311], and Griebel and Oswald [146].

The following results, which also explain the central role of the above splitting concept, are derived from Griebel and Oswald [144, 147]; see Hackbusch [169] for statements of this type in the matrix case.

Lemma 2.2. *(Additive and multiplicative Schwarz iteration). Suppose that V is finite-dimensional, and that the splitting is finite. Let the characteristic numbers $\lambda_{\max}$, $\lambda_{\min}$, and κ of the splitting be given by Definitions (2.12), (2.13). Furthermore, let $\|\tilde{L}\|_{\tilde{V} \to \tilde{V}}$ be the norm of the lower triangular matrix operator $\tilde{L}$ (cf. decomposition 2.9) as an operator in $\tilde{V}$, let $\omega_1 \equiv \lambda_{\max}(\tilde{D}) = \max_{j=0,\dots,J} \left\{ \max_{u_j \in V_j} \frac{a(u_j, u_j)}{b_j(u_j, u_j)} \right\}$ and let $\tilde{W} := \frac{1}{\omega}\tilde{I} + \tilde{L}$.*

(a) *The additive method (2.7) converges for $0 < \omega < 2/\lambda_{\max}$, with the convergence rate*

$$\rho_{as} = \max\{|1 - \omega \cdot \lambda_{\min}|, |1 - \omega \cdot \lambda_{\max}|\} . \qquad (2.15)$$

The bound in (2.15) takes the minimum

$$\rho_{as}^* = 1 - \frac{2}{1 + \kappa} \quad for \quad \omega^* = \frac{2}{\lambda_{\max} + \lambda_{\min}} . \qquad (2.16)$$

(b) *The multiplicative method (2.8) converges for $0 < \omega < 2/\omega_1$, with a bound for the asymptotic convergence rate given by*

$$\rho_{ms} \leq \sqrt{1 - \frac{\lambda_{\min} \cdot (\frac{2}{\omega} - \omega_1)}{\|\tilde{W}\|_{\tilde{V} \to \tilde{V}}^2}} \leq \sqrt{1 - \frac{\lambda_{\min} \cdot (\frac{2}{\omega} - \omega_1)}{(\frac{1}{\omega} + \|\tilde{L}\|_{\tilde{V} \to \tilde{V}})^2}} . \qquad (2.17)$$

The bound in (2.17) takes its minimum

$$\rho_{ms}^* \leq \sqrt{1 - \frac{\lambda_{\min}}{2\|\tilde{L}\|_{\tilde{V} \to \tilde{V}} + \omega_1}} \quad for \quad \omega^* = \frac{1}{\|\tilde{L}\|_{\tilde{V} \to \tilde{V}} + \omega_1} . \qquad (2.18)$$

Proof of Lemma 2.2: the error propagation operator of (2.7) (in V) is $M_{as} \equiv I - \omega \cdot P$ where I denotes the identity in V. Now, for $0 < \omega < 2/\lambda_{\max}$ we have $-1 < 1 - \omega\lambda_{\max} \leq 1 - \omega\lambda_{\min} < 1$ and with (2.15) we obtain $\rho(M_{as}) < 1$, i.e. convergence.

If, on the other hand, we assume convergence, i.e. $\rho(M_{as}) < 1$, we get from $1 > \lambda_{\max}(M_{as}) \geq |1 - \omega\lambda_{\max}| \geq 1 - \omega\lambda_{\max}$ that $\omega\lambda_{\max} > 0$ and consequently $\omega > 0$ since $\lambda_{\max} > 0$ (P has a positive real spectrum). Analogously, the relation $-1 < -\rho(M_{as}) \leq -|1 - \omega\lambda_{\max}| \leq 1 - \omega\lambda_{\max}$ leads to $\omega\lambda_{\max} < 2$, which gives directly $\omega < 2/\lambda_{\max}$.

Now, the optimal value ω^* is just the intersection of the lines $y(\omega) = \omega\lambda_{\max} - 1$ and $z(\omega) = 1 - \omega\lambda_{\min}$. This leads to estimate (2.16). $\square$

Note that using the representation (2.10), the proofs are exactly the same as for the traditional Richardson iteration and the SOR iteration, compare e.g. Hackbusch [169, pp. 82–96]. Note also that, even in the case of divergence, the additive subspace correction serves as an optimal preconditioner as long as κ is independent of J.

Now, we see a difference between the additive and multiplicative method. For the additive method, basically the terms $\lambda_{\min}$ and $\lambda_{\max}$ enter the convergence rate estimates whereas for the multiplicative method the terms $\lambda_{\min}$

and $\|\tilde{L}\|_{\tilde{V}\to\tilde{V}}$ are involved. Note that $\|\tilde{L}\|_{\tilde{V}\to\tilde{V}}$ depends on the ordering of the successive subspace corrections.

Without any additional assumption, there exists an estimate of $\|\tilde{L}\|_{\tilde{V}\to\tilde{V}}$ in terms of $\lambda_{\max}$:

$$\|\tilde{L}\|_{\tilde{V}\to\tilde{V}} \leq \frac{1}{2}[\log_2(2J)] \cdot \lambda_{\max} . \qquad (2.19)$$

This links the estimates of the convergence rate of the multiplicative method to that of the additive method. This estimate is sharp in the general case, for the proof and further details see Oswald [232] and Griebel and Oswald [146].

However, for certain splittings, $\|\tilde{L}\|_{\tilde{V}\to\tilde{V}}$ can be estimated from above by a constant independent of J. In J. Xu [311] and Zhang [318], for example, additional assumptions, where basically strengthened Cauchy-Schwarz inequalities have to be fulfilled, allow to estimate both $\lambda_{\max}$ and $\|\tilde{L}\|_{\tilde{V}\to\tilde{V}}$ from above by the common upper bound $\| |\tilde{P}| \|_{\tilde{V}\to\tilde{V}}$ where $|.|$ denotes element-wise absolute values, i.e.

$$\left.\begin{array}{c}\|\tilde{L}\|_{\tilde{V}\to\tilde{V}}\\[4pt]\lambda_{\max}\end{array}\right\} \leq \| |\tilde{P}| \|_{\tilde{V}\to\tilde{V}} \leq \text{const}.$$

Note that $\| |\tilde{P}| \|_{\tilde{V}\to\tilde{V}}$ is an upper bound for *all* possible $\|\tilde{L}\|_{\tilde{V}\to\tilde{V}}$ that arise for all possible traversal orderings of the multiplicative scheme. Now, the convergence rate of the multiplicative method as well as the additive method is independent of J, if, in addition, the splitting has the property that $\lambda_{\min}$ can be estimated from below by a positive constant which is independent of J.

Now we give an estimate of $\|\tilde{L}\|_{\tilde{V}\to\tilde{V}}$ in terms of $\lambda_{\max}$ without any additional assumption. This links the estimates of the convergence rate of the multiplicative method to that of the additive method.

The general aim is to find splittings $\sum_{j=0}^{J} V_j$ with associated bilinear forms $b_j(.,.)$ such that

- $\lambda_{\min}, \lambda_{\max}, \|\tilde{L}\|_{\tilde{V}\to\tilde{V}}$ or, alternatively, $\| |\tilde{P}| \|_{\tilde{V}\to\tilde{V}}$ can be estimated independently of J,

- the additive or multiplicative method can be implemented using $\mathcal{O}(\dim(V))$ operations per iteration step only and

- hopefully the terms above are independent of the ellipticity constants of the operator, i.e. together with the independence of J, robustness with respect to variations of the coefficients of the PDE is gained.

Note that the first two conditions can be fulfilled by e.g. multigrid or multilevel methods whereas the question of robustness is still not settled in a satisfactory manner.

2.3   Multigrid and Multilevel Methods

Now we have the necessary prerequisites for the analysis of multigrid and domain decomposition methods, which we interpret as subspace correction schemes. The subspace splitting itself determines the type of algorithm, whether it is a multigrid and multilevel method with a nested set of subspaces or it is a domain decomposition method with subspaces constructed on a partition of the domain. After some review of the historical development of multigrid methods, the analysis and different versions of the method are discussed. The main idea is to present the vast number of versions and variants developed so far within a common framework.

The article by Fedorenko [117] is usually considered as the beginning of the history of multigrid methods. Note that for the solution of integral equations Brakhage [56] proposed a two-grid method even earlier. Fedorenko [118] proposed and analyses a two-grid method and later a multigrid method, namely a W-cycle with pre- and post-smoothing using damped Jacobi iterations for the Poisson problem on the unit square. A variable coefficient problem was analysed by Bachvalov [16], structured meshes of triangles' were considered by Astrakhantsev [9] and the multigrid method was further developed to general meshes by Nicolaides [222]. The interest in multigrid methods was mainly theoretical at that time and was focused on the optimal complexity of the algorithms. Real applications were first considered by Brandt [64], who observed the computational efficiency of multigrid methods. He optimised the components of the algorithms in [67] and applied them to adaptively patch-wise refined meshes in [64] and to non-linear problems (FAS scheme, see Brandt [65]). Independently, Hackbusch developed the multigrid method in [163, 164] and applied it to variable coefficient general second order nine-point stencil discretisations on arbitrary domains. The analysis of the multigrid method was continued for finite difference stencils by Wesseling [304] and for more general error norms by Bank and Dupont [19] leading to what is now called the classical multigrid theory. An abstract convergence proof based on an approximation and a smoothing property was established by Hackbusch [165]. The first V-cycle convergence result for the Poisson equation was proved by Braess [52]. Furthermore, the development was summarised in the book Hackbusch [167], see also the conference proceedings series European Multigrid [218] and Copper Mountain [92] and the monographs by McCormick [210], Wesseling [305], Briggs, Henson and McCormick [73], and Trottenberg, Oosterlee and Schuller [295].

As an example, we consider the two-level algorithm. Here, we have two spaces, the fine mesh space V_J of piecewise linear functions on the uniform mesh Ω_J over $\Omega = [0,1]^d$ with mesh size 2^{-J} and the coarse mesh space V_{J-1} of piecewise linear functions on the uniform mesh Ω_{J-1} with mesh size $2^{-(J-1)}$. A mapping between these spaces is given by the so-called prolongation operator $P_{J-1}^J : V_{J-1} \to V_J$ which resembles linear interpolation, and the so-called restriction operator $R_J^{J-1} : V_J \to V_{J-1}$ (e.g. the adjoint of P_{J-1}^J). Discretisation on level J results in the system $A_J u_J = f_J$ and discretisation on level $J-1$ gives the stiffness matrix A_{J-1}. Now an iteration step of the two-level method consists of the following two parts: first, apply ν steps of a so-called smoother, i.e. a classical (convergent) iteration on level J

$$u_J^{it+i/(2\nu)} = u_J^{it+(i-1)/(2\nu)} - C_J(A_J u_J^{it+(i-1)/(2\nu)} - f_J), \quad \text{for } i = 1,\ldots,\nu \quad (2.20)$$

like e.g. a Jacobi iteration, where $C_J = diag(A_J)^{-1}$. Alternatively, also Gauss-Seidel, SOR or many other iterative schemes can be used here. Second, apply a coarse mesh correction step where first the residual is formed and transported (inter-grid transfer operator) to the coarser level by the restriction operator, then the associated coarse level problem is solved exactly and third, using the prolongation operator, this coarse mesh solution is used to update the fine mesh iterate, i.e.

$$u_J^{it+1} = u_J^{it+1/2} - P_{J-1}^J A_{J-1}^{-1} R_J^{J-1}(A_J u_J^{it+1/2} - f_J). \quad (2.21)$$

Of course the two-level method can be used recursively instead of A_{J-1}^{-1} in the coarse mesh correction step (2.21). This results in a general multigrid method. Here a variety of cycling strategies has been developed (V-cycle, W-cycle, compare Hackbusch [167]).

Now, subtracting the exact solution u_J of level J from iteration steps (2.20) and (2.21), using the relation $A_J u_J = f_J$ and plugging (2.20) into (2.21) results in the error equation

$$e_J^{it+1} = (I_J - P_{J-1}^J A_{J-1}^{-1} R_J^{J-1} A_J) \cdot (I_J - C_J A_J)^\nu \cdot e_J^{it}. \quad (2.22)$$

Note that $(I_J - P_{J-1}^J A_{J-1}^{-1} R_J^{J-1} A_J) \cdot (I_J - C_J A_J)^\nu = (A_J^{-1} - P_{J-1}^J A_{J-1}^{-1} R_J^{J-1}) \cdot A_J \cdot (I_J - C_J A_J)^\nu$ holds. Then, using the so-called *smoothing property*

$$\|A_J \cdot (I_J - C_J A_J)^\nu\| \le \eta(\nu)\|A_J\| \qquad \text{for all } \nu \in \mathbb{N}_0 \quad (2.23)$$

with $\lim_{\nu \to \infty} \eta(\nu) = 0$ and the so called *approximation property*

$$\|A_J^{-1} - P_{J-1}^J A_{J-1}^{-1} R_J^{J-1}\| \le C/\|A_J\| \quad (2.24)$$

we get an upper estimate for the convergence rate of the two level method (2.22) by $C/\|A_J\| \cdot \eta(\nu)\|A_J\| = C \cdot \eta(\nu)$, where C and $\eta(\nu)$ are independent of J.

Now, for a given $0 < \xi < 1$ there exists a lower bound $\bar{\nu}$ such that $C \cdot \eta(\nu) < \xi$ for all $\nu \geq \bar{\nu}$. Of course, the validity of the smoothing property and the approximation property has to be shown. This basic approach can be generalised to various settings using different components in the multigrid procedure, for details we refer to Hackbusch [167]. Note that for the approximation property Equation (2.24) to hold, often an additional assumption on the regularity of the continuous problem under consideration, i.e. H^2-regularity, is needed.

Regularity free proofs, proofs for non-uniform meshes and for the additive variant of the multigrid method were obtained differently. Proofs for the hierarchical basis preconditioner and for the hierarchical basis multigrid method were given by Yserentant [314] and Bank, Dupont and Yserentant [20]. Hierarchical basis methods differ from other multilevel methods by the fact, that a node is present just on a single level, namely the coarsest level, whereas in other multilevel methods, the node represent different functions beginning from the coarsest level the node is introduced. However, the hierarchical basis does not lead to optimal methods for dimension $d \geq 2$. Almost regularity free proofs for the additive and the multiplicative multilevel methods were given by J. Xu [310] and Bramble, Pasciak and Xu [63]. However, there, only sub-optimal convergence rates dependent on J could be shown. Then, for the additive methods, the first proof of an optimal $\mathcal{O}(1)$ condition number was given by Oswald [228] without any additional regularity assumptions. Proofs for non-uniform meshes by Dahmen and Kunoth [98] and for multiplicative multigrid by Zhang [318] and Bramble, Pasciak, Wang and Xu [62] followed soon. Further details can be found in [57, 71, 138, 231, 311, 315, 320].

With respect to the convergence theory of section 2.2.2, these multilevel and multigrid methods are based on a nested sequence of subspaces $V_0 \subset V_1 \subset \ldots \subset V_J$, where the spaces V_j are chosen as piecewise linear functions on the corresponding sequence of nested meshes $\Omega_0 \subset \ldots \subset \Omega_J$. In other words, we have a level-wise splitting

$$V = \sum_{j=0}^{J} V_j$$

where the associated auxiliary forms b_j are given by the smoothers on level j. This splitting can be further decomposed into a splitting of one dimensional subspaces

$$V = \sum_{j=0}^{J} V_j = \sum_{j=0}^{J} \sum_{i} V_{j,i} \tag{2.25}$$

where $V_{j,i} = span(\phi_{j,i})$ and $\phi_{j,i}$ denote the usual linear basis functions on mesh Ω_j. Here the auxiliary forms can even be chosen as $b_{j,i} = a_{j,i}$. Now, the additive subspace correction method (2.7) results in the BPX-preconditioner by Bramble, Pasciak and Xu [63] whereas the multiplicative subspace correction method (2.8) is basically equivalent to a multigrid method with Gauss-Seidel pre- or post-smoothing, if a level-wise traversal is chosen. With respect to the convergence theory of section 2.2.2 it can be shown that our multilevel splitting (2.25) results in values for $\lambda_{\min}$, $\lambda_{\max}$, $\|\tilde{L}\|_{\tilde{V}\to\tilde{V}}$ or, alternatively, $\||\tilde{P}|\|_{\tilde{V}\to\tilde{V}}$, which can be estimated independently of J, see also [49, 137, 138, 231, 311, 315, 318].

Thus, in this respect, optimality is achieved. However the constants do depend on the coefficient functions of the underlying operator, i.e. they depend on the ellipticity constants. Therefore, in practice the convergence rate is still dependent on the coefficient functions of the underlying operator. Now the question arises whether there is a multilevel method with a convergence rate independent of the coefficients, which is *robust* in this respect. To achieve robustness, various modifications of the conventional multigrid scheme have been introduced: first of all, modifications of the smoother have been proposed, i.e. line- or zebra-smoothers by Stüben and Trottenberg [289] or ILU-type smoothing procedures by Wittum [309] and Stevenson [287]. Furthermore, instead of modifying the smoother, the coarse mesh spaces also can be modified by using semi-coarsening, matrix dependent prolongations and restrictions by De Zeeuw [316], Fuhrmann [121], and Reusken [251, 252] or even specific algebraic coarsening techniques, i.e. so called algebraic multigrid methods, see Brandt [68], Ruge and Stüben [259], Axelsson and Vassilevski [10], and Grauschopf, Griebel and Regler [131]. For the case of elasticity we refer to Braess [53] and Vanek, Mandel and Brezina [297]. Further modifications have been proposed to deal with complex geometries, where nested hierarchies of meshes are not available, see Kornhuber and Yserentant [189], Bank and Xu [23], and Hackbusch and Sauter [170] in addition to the algebraic multigrid methods. These approaches work well and are apparently robust in practical experiments. However, a proof of robustness for algebraic multigrid and similar variants does not exist. For simple separable self-adjoint operators, however, a robust additive multilevel method based on pre-wavelets has been presented and proved rigorously by Griebel and Oswald [147]. Note however that for the general case in three dimensions there is still the question of how a robust multigrid method could be constructed and how robustness can be proved.

Note that multigrid methods have been applied to problems in elasticity using conforming, non-conforming and mixed discretisations in [176] of different plate bending models in [229, 239, 319], shell models in [235] and for plane elasticity in [29, 55, 70, 196, 322]. However, in the case of elasticity applications, the robustness of a multigrid method is also subject to actual research. Here the convergence rates are of course independent of the number of unknowns, however, they depend on anisotropies of the material constants, on the aspect ratio of the domain Ω, on the Poisson number and on locking phenomena of a discretisation.

Note that non-linear equations can be treated using the multigrid approach as well: either a Newton iteration can be used as an outer iteration with a linear multigrid method or alternatively nonlinear multigrid versions (Brandt's FAS [65] and Hackbusch's nonlinear multigrid [167]) with Picard or Newton-type smoothers, for a comparison see Hemker and Koren [173]. Further non-linear subspace corrections methods including theory can be found in J. Xu [312], Tai and Espedal [290], and Tai and Xu [291].

2.4   Domain Decomposition Methods

Closely related to the highly efficient multigrid methods are domain decomposition methods. They are related by the common analysis framework of subspace correction methods. Furthermore, the aim of the methods is similar, namely the efficient solution of a linear equation system, which arises in the discretisation of a partial differential equation. However, the approach of domain decomposition methods is different: conceptually the parallelism of the methods is stressed. The task of solving the large equation system is broken up into several independent smaller tasks of solving smaller equation systems. A geometrical decomposition of the domain instead of a level-wise decomposition as for multigrid methods is used, which in turn leads to a slightly worse complexity estimate.

The application of the well known *divide et impera* principle leads in a natural way to domain decomposition techniques. Here the domain is partitioned into subdomains on which the small problems which arise can be solved easily. Then the local solutions have to be combined and processed further to obtain the solution of the overall problem. The partitioning can be given in a natural way, e.g. by the geometry or the material coefficients of the problem. It can also be heuristically constructed by mesh partitioning techniques. In addition a domain decomposition is advantageous for later parallelisation. Based on this principle a series of algorithms has been developed, i.e. overlapping Schwarz methods, non-overlapping Schur complement methods, etc. The development of these methods up to now can be found in the conference proceedings 'domain decomposition methods' [104], see also the surveys by Chan and Mathew [85], Smith, Bjørstad and Gropp [278], and Xu and Zou [313].

2.4.1 Overlapping Schwarz methods

The first domain decomposition methods for the solution of PDEs were the overlapping Schwarz methods, named after H. A. Schwarz [270]. They were used for the existence proof of a continuous solution of the Poisson equation in a domain, which was composed of the union of two overlapping standard domains, see Sobolev [282], Morgenstern [217], Babuška [12], and Lions [202] for further early references. In the general case, the basic idea is to construct the domain Ω as the union of several subdomains $\Omega = \bigcup_j \Omega_j$ with overlap $\Omega_j \cap \bigcup_{i \neq j} \Omega_i \neq \emptyset$ of positive diameter along inner boundaries $\partial \Omega_j \setminus \partial \Omega$. Discretisations on the overlapping domains usually are based on matching meshes: the elements or mesh cells of two subdomains Ω_i and Ω_j coincide within the overlap $\Omega_i \cap \Omega_j$. Thus the overlap has to be at least one element wide. The corresponding spaces $\hat{V}_j$ are constructed by a discretisation on the subdomain Ω_j with Dirichlet boundary conditions on the inner boundary $\partial \Omega_j \setminus \partial \Omega$ along with the bilinear form $a(.,.)$ restricted to Ω_j. The splitting looks like

$$V = \sum_j^J \hat{V}_j ,$$

where the associated auxiliary forms b_j are given by exact or inexact subdomain solvers on Ω_j. The associated iterative methods (2.7) and (2.8) lead to the additive and multiplicative Schwarz method, which can be analysed with the convergence theory of section 2.2.2. The convergence rate depends on the size of the overlap and on the number of subdomains, see Table 2.2.

Furthermore, the introduction of a coarse mesh removes the latter dependence and facilitates immediate information transport between pairs of non-overlapping subdomains, in analogy to the two-level multigrid method. For example, for a Poisson problem a coarse mesh with one degree of freedom per subdomain is sufficient for this purpose. The coarse space V_{coarse} can be chosen as the span of piecewise constant or piecewise linear functions on the subdomain scale, where each function is associated with one subdomain. We denote the mesh size of the coarse mesh by H and denote the size of the overlap by $\beta \cdot H$ to indicate that the overlap should be chosen independently of the fine mesh size h. Now, the splitting is

$$V = V_{\text{coarse}} + \sum_{j=0}^J \hat{V}_j ,$$

where the new associated form is usually chosen as $b_{\text{coarse}}(.,.) = a_{\text{coarse}}$, i.e. the restriction of the bilinear form $a(.,.)$ to V_{coarse}. This approach results in a

Algorithm	Convergence rate	Subspace splitting
Overlapping Schwarz iteration		
additive Schwarz	$CH^{-2}(1+1/\beta^2)$	$V = \sum_j \hat{V}_j$
— with coarse mesh	$C(1+1/\beta^2)$	$V = V_{\text{coarse}} + \sum_j \hat{V}_j$
Schur complement iteration		
nodal basis	$CH^{-1}h^{-1}$	$V = V_\Gamma + \sum_j V_j$
hierarchical basis $d=2$	$C(1+\log^2(H/h))$	$V = V_\Gamma^{\text{hb}} + \sum_j V_j$
$d=3$	$CH/h(1+\log^2(H/h))$	$V = V_\Gamma^{\text{hb}} + \sum_j V_j$
multilevel basis	C	$V = V_\Gamma^{\text{BPX}} + \sum_j V_j$
Preconditioner for the Schur complement iteration		
BPS $\quad\quad d=2$	$C(1+\log^2(H/h))$	$V_\Gamma = V_{\text{coarse}} + \sum_k V_{\text{edge } k}$
wire-basket $\quad d=3$	$C(1+\log^2(H/h))$	$V_\Gamma = V_{\text{coarse}} + \sum_k V_{\text{edge } k}$ $+ \sum_l V_{\text{face } l}$
vertex space $\quad d=2$	$C(1+\log^2\beta)$	$V_\Gamma = V_{\text{coarse}} + \sum_k V_{\text{edge } k}$ $+ \sum_m V_{\text{vertex } m}$
$d=3$	$C(1+\log^2\beta)$	$V_\Gamma = V_{\text{coarse}} + \sum_k V_{\text{edge } k}$ $+ \sum_l V_{\text{face } l}$ $+ \sum_m V_{\text{vertex } m}$
Neumann-Neumann	$CH^{-2}(1+\log^2(H/h))$	$V_\Gamma = \sum_j V_j^{\text{Neum.}}$
— with coarse mesh	$C(1+\log^2(H/h))$	$V_\Gamma = V_{\text{coarse}} + \sum_j V_j^{\text{Neum.}}$

Table 2.2. *Convergence rates of domain decomposition methods for constant coefficient problems with a coarse mesh of mesh size H and a fine mesh mesh size h. The value βH denotes the mesh overlap for the overlapping Schwarz method and the diameter of the vertex spaces for the vertex space method, respectively.*

convergence rate which is independent of h. However, overlapping methods in general are not robust with respect to discontinuous coefficients.

2.4.2   Non-overlapping Schur complement methods

The history of non-overlapping domain decomposition methods begins with direct substructuring methods. The domain Ω is partitioned into two disjoint subdomains Ω_1 and Ω_2. Then, the separator $\Gamma = \partial\Omega_1 \setminus \partial\Omega$ is removed and the two local sub-problems are solved before we solve for the unknowns on the separator. This scheme can be applied recursively to the sub-problems and

Algorithm		Convergence rate
additive Schwarz	$d = 2$	$C_\beta(1 + \log(H/h))$
	$d = 3$	$C_\beta H/h$
vertex space	$d = 2$	$C_\beta(1 + \log(H/h))$
	$d = 3$	$C_\beta H/h$
BPS	$d = 2$	$C(1 + \log^2(H/h))$
wire-basket	$d = 3$	$C(1 + \log^2(H/h))$
Neumann-Neumann	$d = 2, 3$	$C(1 + \log^2(H/h))$

Table 2.3. *Convergence rates of domain decomposition methods for discontinuous coefficients. Discontinuities are aligned to domain boundaries. Constants depend on the mesh overlap βH for the additive Schwarz and the vertex space method.*

is then equivalent to a Gaussian elimination with nested dissection ordering. Early references for this method, which is popular in structural mechanics, can be found in Przemieniecki [247]. In case of the simple partitioning into two subdomains the stiffness matrix is block partitioned correspondingly, i.e.

$$A = \begin{pmatrix} A_{\Gamma\Gamma} & A_{\Gamma 1} & A_{\Gamma 2} \\ A_{1\Gamma} & A_{11} & 0 \\ A_{2\Gamma} & 0 & A_{22} \end{pmatrix} .$$

Now, iterative substructuring methods are based on the iterative solution of the Schur complement system $S = A_{\Gamma\Gamma} - \sum_{i=1}^{2} A_{\Gamma i} A_{ii}^{-1} A_{i\Gamma}$ for the unknowns on the separator Γ. In some algorithms the subdomain problems A_{ii}^{-1} are also solved iteratively.

The two domain case can easily be generalised to the many domain case. Here Ω is partitioned into a union of disjoint subdomains $\Omega = \bigcup_j \overline{\Omega_j}$ without overlap $\Omega_j \cap \Omega_i = \emptyset$ for $i \neq j$. For the Ω_j we use meshes which match at the interface

$$\Gamma = \bigcup_j \partial\Omega_j \setminus \partial\Omega .$$

The discretisations in the subdomains Ω_j also have to match. Otherwise, we have to use more sophisticated methods. In addition to the degrees of freedom in the subdomains, there are degrees of freedom on the interface Γ. Now, the space V_j consists of functions which vanish outside the open domain Ω_j. The interface space V_Γ is defined as the span of shape functions of the interface mesh Ω_Γ, which are extended to the adjacent subdomains Ω_J in some way and vanish on the boundary $\partial\Omega$. Then the subspace splitting according to the convergence theory of section 2.2.2 looks like

$$V = V_\Gamma + \sum_{j=0}^{J} V_j\,.$$

We choose the associated auxiliary forms as $b_j(.,.) = a_j(.,.)$ and use a multi-plicative Schwarz method. The choice of the Poincaré-Steklov operator $b_\Gamma(x,y) \equiv x^T S y$ with exact solvers would lead to the direct substructuring method again. However, the Schur complement is expensive to compute and is not directly accessible in an iterative procedure. Hence we choose the auxiliary form on the separator as the L_2 scalar product $b_\Gamma(.,.) = (.,.)$ along with one Richardson iteration step. A conjugate gradient method in V with this pre-conditioner is equivalent to a conjugate gradient method applied to the smaller Schur complement system on V_Γ. The condition number of this preconditioner is $CH^{-1}h^{-1}$ with a mesh size h of Ω and a subdomain diameter of H, if as a basis for V_Γ just the nodal basis is chosen, see also Table 2.2.

In general there are two ways to improve the condition number further: we can extend the functions of V_Γ in a more clever way by the hierarchical basis Smith and Widlund [279] and Haase, Langer and Meyer [160] or a multilevel technique Tong, Chan and Kuo [294], Griebel [139], and Haase, Langer, Meyer and Nepomnyaschikh [162]. This results more or less in a harmonic extension operator, which couples the interior of the subdomains and the separator, see also Table 2.2. Alternatively we can construct preconditioners for the interface Schur complement, which results in improved forms $b_\Gamma(.,.)$, which we will consider in the next section. Further modifications include the use of approximative solvers $b_j(.,.)$ on the subdomains, see Börgers [47] and Haase, Langer and Meyer [161].

2.4.3 Preconditioners for the Schur complement

We are interested in finding a preconditioner for the interface problem in the space V_Γ. Such a preconditioner can be used in an iteration on the Schur com-plement S or in the construction of a global preconditioner for the operator A. Historically, the first domain decomposition preconditioners for the Schur com-plement were designed for the two domain case only. Here, implementations of the analytic preconditioner for $H^{1/2}(\Gamma)$ by Dryja [108] or the improved version by Golub and Mayers [129] were based on the discrete sine transform which is a real number version of the Fast Fourier Transform (FFT) [246]. The next step was to construct preconditioners for the case of many subdomains. Then crosspoints were contained in the separator. Bramble, Pasciak and Schatz were able to treat this case with crosspoints [59]. This algorithm and its three dimensional generalisation, the wire-basket preconditioner by Bramble, Pasciak and Schatz [60] and Smith [277] are based on a splitting of the space

$$
\begin{aligned}
V_\Gamma &= V_{\text{coarse}} + \sum_k V_{\text{edge } k} && \text{for } d = 2, \text{ and} \\
V_\Gamma &= V_{\text{coarse}} + \sum_k V_{\text{edge } k} + \sum_l V_{\text{face } l} && \text{for } d = 3, \text{ respectively,}
\end{aligned}
\tag{2.26}
$$

into one global coarse mesh space of cross points and into local spaces for each edge or face $\subset \partial\Omega_i \setminus \partial\Omega \subset \Gamma$, which connects crosspoints or crosspoints with the boundary $\partial\Omega$. The associated bilinear form of the coarse mesh is $b_{\text{coarse}}(.,.) = a_{\text{coarse}}(.,.)$ and the coarse system is solved directly. However, the bilinear forms of the edges $b_{\text{edge } k}$ (and the faces $b_{\text{face } l}$) are not available, similar to the Schur complement itself. Hence other Schur complement preconditioners have to be employed for the spaces $V_{\text{edge } k}$ and $V_{\text{face } l}$, such as the above-mentioned preconditioner of Dryja. The resulting convergence rates can be found in Table 2.2.

An extension of this construction principle leads to the vertex space method by Smith [276]. It is related to the fictitious domain methods by Nepomnyaschikh [207, 219, 220] and Agoshkov and Lebedev [1]. In addition to the splitting in Equation (2.26), for each vertex a space is constructed which consists of functions on the separator Γ in the vicinity of the vertex x_m, i.e.

$$
\Omega_{\text{vertex } m} = \mathbb{B}_{\beta H}(x_m) \cap \Gamma ,
$$

where $\mathbb{B}_{\beta H}(x_m)$ denotes the ball around x_m with radius βH.

A different class of Schur complement preconditioners is based on the solution of auxiliary problems on the subdomains. The Neumann-Dirichlet preconditioner by Bjørstad and Widlund [45] is based on the solution of a Neumann problem on half of the subdomains, and in a second step, on Dirichlet problems on the remaining subdomains. A popular variant of it is the Neumann-Neumann preconditioner , see Bourgat, Glowinski, LeTallec and Vidrascu [51] and LeTallec, de Roeck and Vidrascu [194]. There, on each subdomain a local Neumann problem is solved. Then the actual right hand side in V_Γ is transformed into a Neumann boundary condition. Finally, the values on $\partial\Omega_j \cap \Gamma$ are transformed back and summed up. This approach can be interpreted in our additive Schwarz framework. We end up with a splitting of the type

$$
V_\Gamma = \sum_j V_j^{\text{Neumann}}
$$

with associated bilinear forms $b_j(.,.) = a_j(.,.)$ with appropriate boundary conditions incorporated. There are variants with a coarse mesh to improve the performance in the many subdomain case, see Mandel [206], LeTallec [193], and Dryja and Widlund [112]:

$$V_\Gamma = V_{\text{coarse}} + \sum_j V_j^{\text{Neumann}}.$$

More recent developments are concerned with the development of analogous preconditioners for discretisations, where the continuity between subdomains Ω_j is maintained through Lagrange multipliers, called FETI method by Farhat and Roux [116] and Mortar method by Bernadi, Maday and Patera [37] and Maday, Mavriplis and Patera [205]. The basic advantage of non-overlapping domain decomposition is its robustness with respect to discontinuous coefficients aligned to subdomain boundaries. In many large scale problems it is possible to partition the domain Ω into subdomains of constant or slowly varying coefficients so that each subdomain consists of exactly one material. Based on this decomposition, the resulting preconditioner is often competitive to other methods with better theoretical properties.

2.5 Sparse Grid Solvers

So far, we have discussed the solution of linear equation systems which arise from the discretisation of partial differential equations by finite element or finite difference methods. We ended up with optimal order solvers of multigrid and domain decomposition type. Of course, there are alternative ways of discretisation of PDEs, e.g. by spectral and higher order methods, which require slightly different solvers. Or to put it in other way: there are discretisation schemes which lead to stiffness matrices that are not sparsely populated. Hence we cannot expect to solve such an equation system in linear order of the number of equations. An optimum could be a linear complexity in the number of non-zero matrix entries. Nevertheless, such discretisations can be competitive with standard discretisations and optimal order solvers, if the solution can be approximated by far fewer degrees of freedom by the discretisation. This is the case for sparse grid discretisations. Since this chapter covers the solution of equation systems, we also have to cope with the problem of solving PDEs on sparse grids. Again, we review the different discretisation schemes available, namely the combination technique, the Galerkin method and the finite difference method. In contrast to the previous sections, we do not present a complete theory, since both results for approximation of functions and the convergence of iterative solvers are not fully developed yet.

Sparse grid schemes are multi-dimensional approximation schemes which are known under several names such as 'hyperbolic crosspoints', 'splitting extrapolation' or as a Boolean sum of grids. Probably Babenko [11] and Smolyak [280] were the first historically references. Directly related to the Boolean construction of the grids was the construction of a multi-dimensional quadrature formula. Both quadrature formulae and the approximation properties of such tensor product spaces were subject to further research, see Zung [325, 326], Temlyakov [292, 293], and others. A Galerkin method was proposed by Cavendish, Gordon and Hall [84]. The *curse* of dimension was also subject to general research on the theoretical complexity of higher-dimensional problems. Compared to regular, uniform meshes of a mesh parameter h which contain h^{-d} points in d dimensions, sparse grids require only $h^{-1}|\log h|^{d-1}$ points due to a truncated, tensor-product multi-scale basis representation. For such reasons, sparse grids play an important role for higher-dimensional problems. Besides the application to quadrature problems, sparse grids are now also used for the solution of PDEs. They provide an efficient approximation method of smooth functions, especially in higher dimensions $d \geq 3$.

Sparse grids were introduced by Zenger [317] for the solution of elliptic partial differential equations, where a Galerkin method, adaptive mesh refinement and tree data structures were discussed. At the same time a different discretisation scheme based on the extrapolation of solutions on several related, regular grids was proposed, see Griebel [135, 136] and Griebel, Schneider and Zenger [149]. So far, Galerkin methods, see the references in Bungartz [79], Bungartz and Dornseifer [80], Bungartz, Dornseifer and Zenger [81], and Schwab and Todor [269], finite difference schemes, see Griebel [140] and Schiekofer [266], and extrapolation methods for elliptic problems on sparse grids have been investigated.

2.5.1 Combination Technique Solvers

Probably the simplest way to solve a PDE on a sparse grid is to use the extrapolation method, also called the *combination-* (Griebel, Schneider and Zenger [149]) or *splitting extrapolation-*technique (Liem, Lu and Shih [201]). The idea is to combine solutions computed on several different regular grids to a more accurate sparse grid solution. This approach uses the concept of extrapolation from numerical analysis to cancel out some low order error terms by the combination of solutions and to achieve a smaller discretisation error.

Let us assume that we have a standard PDE solver for regular grids of $n_1 \cdot n_2 \cdot \ldots \cdot n_d$ nodes. This can be any software and solver capable of uniform grids on rectangular shaped domains or anisotropic grids on the unit square respectively. Let us denote such an anisotropic grid as $G_{h_1,h_2,...,h_d}$. Furthermore the mesh parameters h_i will always be of the form $h_i = 2^{-j_i}$ with a multi-index j. A sparse grid of level l can be decomposed into the sum of several regular grids

$$G_l^{\text{sparse}} = \bigcup_{|j|=l} G_j \,,$$

see also Figure 2.5. The idea is now to use this decomposition and the numerical solution of the PDE and to decompose the solution u into solutions u_j on regular grids. This can be done in the two-dimensional case as

$$u_l^{\text{sparse}} := \sum_{|j|=l} u_j - \sum_{|j|=l-1} u_j \,, \tag{2.27}$$

which is an extrapolation formula with weights $+1$ and -1. The general d-dimensional expression is

$$u_l^{\text{sparse}} := \sum_{i=0}^{d-1} (-1)^i \binom{d-1}{i} \sum_{|j|=l-i} u_j \,. \tag{2.28}$$

If we depict the grid points of a grid G_j, each node is associated with a shape function in a FEM discretisation. The solution originally obtained only in the nodes is extended by the shape functions of a linear FEM discretisation to a piecewise (multi-) linear function on the whole domain. Now we can sum up functions from several grids e.g. $G_{2,0}$, $G_{1,1}$ and $G_{0,2}$. The union of the nodes gives the nodes of the corresponding sparse grid, compare Figures 2.4 and 2.5. Computationally, the combination of the grids requires interpolation on each grid, or the evaluation of the shape functions on the sparse grid points, because the sparse grid always contains a superset of nodes of each of the regular grids. The remaining nodes of a regular grid are determined by interpolation. For the correct extrapolation we also need to subtract several coarse grids, which can be done in the same way.

The fact that the extrapolation solution equals the solution of the PDE on the sparse grid can be proven by an error expansion of the regular grid solutions and the cancellation of the lower order error terms, see Griebel, Schneider and Zenger [149]. The quality of an extrapolation formula, i.e. the sum of the remaining error terms, depends on higher derivatives of the solution u and can be measured in higher order Sobolev norms (e.g. $H^{2d}(\Omega)$). Given sufficient smoothness, it is possible to combine solutions obtained by Finite Elements, Finite Volumes or Finite Differences. The latter two require an appropriate, order-preserving interpolation scheme for the nodal values obtained on single

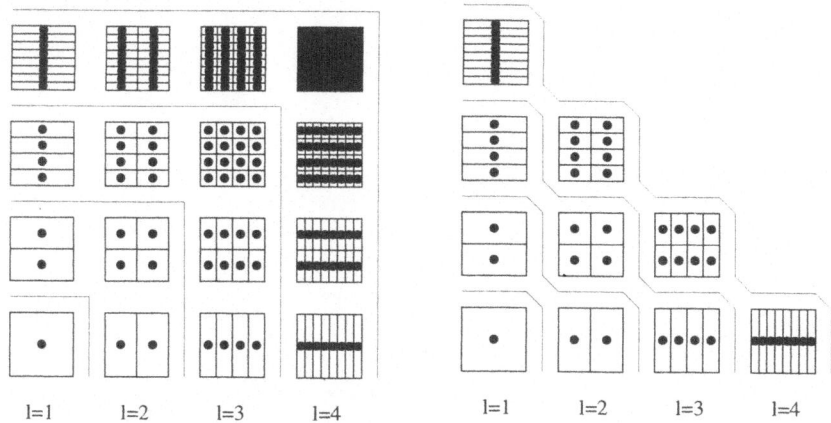

Figure 2.3. *Tableau of supports of the hierarchical basis functions spanning a two-dimensional regular space (left) and the corresponding sparse grid space.*

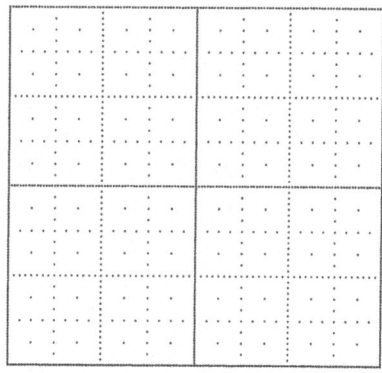

Figure 2.4. *A sparse grid on the unit square. Depicted are the nodes associated to the hierarchic basis functions.*

grids. The FEM provides a natural extension of the solution from the nodes to the whole domain through the shape functions used. Error bounds in different norm are available and the error usually compares to the regular grid error, with a logarithmic deterioration $|\log h|$.

Besides its simplicity, there are several advantages of this method: the solution can be computed concurrently, that is on different processors of a parallel computer, which is almost an embarrassingly parallel algorithm, see also Griebel [135], or one after the other requiring just a little computer mem-

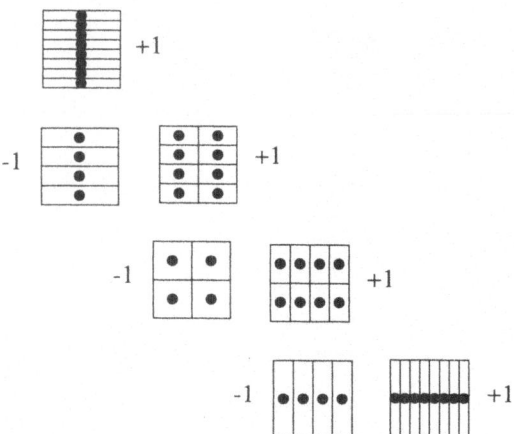

Figure 2.5. *The sparse grid combination technique. Several solutions of the PDE obtained on different regular grids by standard PDE codes are summed up to form a sparse grid solution of higher accuracy. The weights of $+1$ and -1 are depicted next to the grids. The grids are arranged analogous to the tableau in Figure 2.3. The union of nodes of all regular grids results in the nodes of the corresponding sparse grid, see Figure 2.4.*

ory. Multiple instances of a standard PDE code can be used and just a little coding is necessary to implement the method. Furthermore such a code may be a very efficient code optimised and tuned for structured grids on a specific computer. The prerequisites of such a code are an interpolation procedure, which defines the solution in the whole domain, and a standard solver, which is able to deal with anisotropic discretisations i.e. a grid of $n_1 \cdot n_2 \cdot \ldots \cdot n_d$ nodes with different n_i.

The convergence of the extrapolation method has been analysed for classical solutions by Bungartz, Griebel, Roschke and Zenger [82] and for usual weak solutions by Pflaum and Zhou [242]. However, the solution has to fulfil some higher order smoothness for a good sparse grid approximation. We denote a mixed derivative tensor-product type of Sobolev space by $\hat{H}^1$, which we define as

$$\hat{H}^1 \;=\; \{u \mid D^i u \in H^1 \;\; \forall |i| \le 1\}$$

with multi-index $i \in \mathbb{N}_0^d$. The corresponding norm of the space reads as

$$\|u\|_{\hat{H}^1}^2 \;=\; \sum_{|i| \le 1} \|D^i u\|_2^2.$$

Note that the norm can be considered as a tensor-product H^1-norm, while

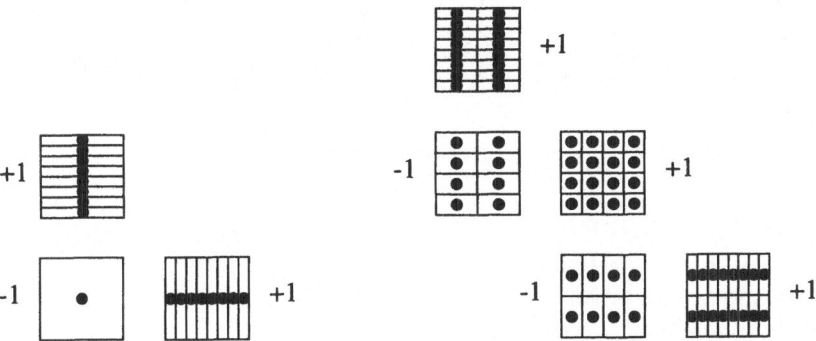

Figure 2.6. *Sparse grid extrapolation schemes: splitting extrapolation (left), 'semi'-sparse grid (right). A smaller number of solutions than for the sparse grid extrapolation technique are combined to form a higher accuracy solution. Again, each solution on a regular grid can be obtained by any suitable software package or PDE solver.*

the space $\hat{H}^1$ is much more like an H^2 and is only slightly larger than the tensor-product H^2 space. A solution $u \in \hat{H}^1$ can be approximated by the combination technique measured in the energy-norm like

$$|||u - u^{\text{sparse}}||| \ \leq \ Ch|\log h|^{d-1} \, \|u\|_{\hat{H}^1} \, . \tag{2.29}$$

The efficient solution of partial differential equations on sparse grids can be done straightforwardly. In contrast to discretisation techniques directly on sparse grids, several independent sub-problems have to be solved for the combination technique. Each sub-problem consists in the solution of the PDE on a regular grid. Direct solvers and several types of iterative solvers can be used here, for example semi-coarsening versions of a multigrid solver. Hence the methods for the solution of large equation systems discussed so far are applicable. Given the complexity of a discretisation and solution on a regular grid, the sparse grid extrapolation method requires just the sum of the single grid complexities and the interpolation procedure. Let us assume that we have an $\mathcal{O}(n)$ linear complexity linear algebra method at hand for regular grid problems of n degrees of freedom. The combination technique sums up in this case to an overall $\mathcal{O}(n)$ method for n degrees of freedom on the sparse grid, which is optimal up to a constant factor. However, more expensive solvers can also be used on sub-problems with $\mathcal{O}(f(n))$, which leads to competitive overall complexity below $\mathcal{O}(f(n))$ for an n degrees of freedom sparse grid. In a similar way the combination technique can be extended to time-dependent and to nonlinear problems, where again several independent solutions are combined in an optimal order solution procedure.

2.5.2 Galerkin Sparse Grid Solvers

The standard approach to discretise an elliptic differential equation on a sparse grid or even on an adaptively refined sparse grid is the Galerkin method. The Galerkin method is also the base of the finite element method (FEM). The main difference between the FEM and sparse grids is that some sparse grid basis functions have global support and are not *finite* as the FEM shape functions with finite size support. Historically, the Galerkin method was first applied to sparse grids by Cavendish, Gordon and Hall [84] and independently by Zenger [317].

Given a grid along with basis functions ϕ_i on that grid, it is straightforward to apply a Galerkin scheme with basis and trial functions ϕ_i. The equation system is derived from the variational form $a(.,.)$ of the partial differential equation: *find $u \in V$ such that*

$$a(u,v) \;=\; f(v) \quad \forall v \in V \,, \tag{2.30}$$

$$\begin{aligned}
\sum_j a_{i,j} u_j &\;=\; f_i \quad \forall i \\
\text{with} \qquad a_{i,j} &\;=\; a(\phi_i, \phi_j) \\
\text{and} \qquad f_i &\;=\; f(\phi_i)
\end{aligned} \tag{2.31}$$

This method can be applied to any set of linear independent basis functions in a finite-dimensional setting. Furthermore, the Galerkin method can be applied to any complete, linear independent basis in the case of infinite-dimensional function spaces like H^1. We define these functions on sparse grids. The functions can be chosen from the hierarchical basis or some other multi-resolution system, so that each function is related to a node of the grid. The idea now is to use shape functions of the sparse grid, which are direct products of one-dimensional functions of the hierarchical basis or some other (pre-) wavelet basis. At a coordinate position x in the domain $\Omega \subset \mathbb{R}^d$ a shape function can be defined by

$$\phi_i(x) = \prod_{j=0}^{d-1} \phi_{i_j}(x_j) \,.$$

In the case of the hierarchical basis, the functions ϕ_i are the standard multi-dimensional hat functions also used in FEM on a rectangle, cube or hypercube shaped elements. The discretisation is symmetric as long as the bilinear form $a(.,.)$ is self-adjoint. The discretisation error can be bounded by the interpolation error of the sparse grid. Error bounds for the energy norm of the Laplacian are available and comparable to the regular grid error, with a logarithmic deterioration $|\log h|$, analogous to the combination technique, see Equation (2.29). However, sharp L_2-norm estimates are not known yet for the Galerkin solution. While best n-term approximation result on sparse grids in L_2 norm does not show the logarithmic deterioration $|\log h|^{d-1}$, see Oswald [233], the L_2 error of the Galerkin solution seems to contain the factor, which was observed computationally, see e.g. Bungartz [78].

There are two main drawbacks to a naive implementation of this approach: the stiffness-matrix $a_{i,j}$ is not sparse, unlike in the FEM case. This is basically due to the fact that many shape functions do have a large support. For the FEM discretisation, these supports are small and only a bounded number of shape functions interact in the computation of $a(\phi_i, \phi_j)$. For the same reason the computation of the right hand side f_j is quite expensive. As a consequence, the performance of the sparse grid degrades almost to the performance of a full h^{-d} grid.

As a consequence, the stiffness matrix is never assembled in an implementation, but an algorithm for a matrix multiply or Gauss-Seidel step on the fly is used. This algorithm can be formulated in terms of tree traversals and indeed has linear complexity, see Zenger [317]. Any method which uses a bounded number of matrix multiplies or Gauss-Seidel iterations has optimal complexity. This may be the case, for example, for some suitable preconditioners or accelerators for the solution of the linear equation system. Extensions of the optimal order sparse grid tree traversal algorithm to several dimensions, see Balder [17], and to certain types of variable coefficients have been developed, see Bungartz, Dornseifer and Zenger [81] and references therein. The main difficulty in this respect is the symmetry of the discretisation. The optimal order algorithms are usually based on the assumption of constant coefficients. The treatment of the variable coefficients case is much more involved. In order to maintain optimal complexity, the coefficient function has to be approximated on the sparse grid. This approximation can cause asymmetries in the operator for a naive implementation. Further difficulties come from jumping coefficients not aligned to a grid axis and from more complicated differential operators.

When it comes to the solution of the linear equation system, there are different methods available, usually interwoven with the matrix multiplication algorithm on the sparse grid. While direct solvers used by Cavendish, Gordon and Hall [84] and ordinary Gauss-Seidel iteration by Zenger [317] are not advisable for large scale problems, some multigrid-like solvers have been developed so far. The multiplicative multigrid method proposed by Griebel [133, 134] and similarly by Pflaum [241] experimentally shows an optimal $\mathcal{O}(1)$ condition number independent of the number of unknowns. Theoretical bounds are currently weaker. Each multigrid cycle requires $\mathcal{O}(n)$ operations for an n node sparse grid. The method can be interpreted as a sequence of several multigrid V-cycles with semi-coarsening along one direction. The main difficulty is the management of a consistent residual, since a direct computation from the fine grid is prohibitively expensive and local updates involve contributions from several grids at a time with semi-coarsening or semi-refinement along different axes. Even more involved is the treatment of variable coefficient differential operators. The coefficient functions will be evaluated only on sparse grid nodes for complexity reasons, which might not be accurate enough e.g. for jumping coefficient operators. Since then the discretisation on sparse grids is complicated, the computation of residuals is even more complicated. However, for the efficient solution of the equation system with smooth coefficients it is often sufficient to use a constant coefficient approximation of the differential operator, and the constant coefficient version of the multiplicative method is applicable.

Theoretically, results for additive multigrid methods on sparse grids have been derived. On standard discretisations, additive multigrid solvers like the BPX-preconditioner can be defined as an additive V-cycle with a single Jacobi step or equivalently diagonal scaling. The generalisation to sparse grids can be based on a sequence of nested regular grids as in the multigrid method for regular grids, which are all subsets of the sparse grid, and additionally a set of anisotropic regular grids needed to cover the remaining nodes of the sparse grid. Given a symmetric discretisation, the preconditioner leads to a condition number of $\mathcal{O}(j^{d-2})$ with the number of levels j, see Griebel and Oswald [145]. This is optimal for the two-dimensional case and deteriorates for large dimensions d. Furthermore, stable wavelet multi-resolution schemes can be used instead of the hierarchical basis. In the case of an H^1-orthogonal basis the diagonal scaling leads to the optimal result of an $\mathcal{O}(1)$ condition number, see Griebel and Oswald [147]. The stability of the basis allows the deletion of arbitrary nodes from a regular grid to arrive at a sparse grid without changing the condition number estimate. Further details on wavelet basis sparse grids can be found in Knapek [185] and Hochmuth, Knapek and Zumbusch [177].

2.5.3 Finite Difference Sparse Grid Solvers

The combination technique discretisation on sparse grids is not applicable to adaptively refined meshes, and the Galerkin discretisation is difficult to implement for varying coefficients and distorted domains. Hence an alternative type of discretisation is sought. Standard finite difference stencils can be used to discretise partial differential equations on arbitrary sparse grids. However, instead of the direct application of ordinary multi-dimensional stencils, the hierarchical basis or wavelet basis has to be used for the interaction between coarse and fine grid basis functions. One way to do this is to use one-dimensional finite difference stencils $\mathbf{D}$ along one coordinate axis and a (pre-) wavelet transformation $\mathbf{B}$ and back-transformation $\mathbf{B}^{-1}$ along the remaining coordinate axes. The transformation changes the basis representation between nodal basis and wavelet basis, both restricted to the sparse grid. The difference stencils $\mathbf{D}_{ii}$ are applied to a node and its geometrically nearest neighbour nodes along the axis, while the transformation leads to the interaction of all scales. The exact term proposed by Griebel [140] and Schiekofer [266] reads as

$$\frac{\partial}{\partial x_i} u \approx \mathbf{B}_{I\setminus\{i\}}^{-1} \circ \mathbf{D}_i \circ \mathbf{B}_{I\setminus\{i\}} u \tag{2.32}$$

$$\frac{\partial^2}{\partial x_i^2} u \approx \mathbf{B}_{I\setminus\{i\}}^{-1} \circ \mathbf{D}_{ii} \circ \mathbf{B}_{I\setminus\{i\}} u \tag{2.33}$$

$$\Delta u \approx \sum_{i=1}^{d} \mathbf{B}_{I\setminus\{i\}}^{-1} \circ \mathbf{D}_{ii} \circ \mathbf{B}_{I\setminus\{i\}} u , \tag{2.34}$$

which is also illustrated in Figure 2.7. A general difference operator is then obtained by dimensional splitting. A linear convection-diffusion equation, as a simple example, can be discretised in nodal basis representation as usual where the one-dimensional difference operators $\mathbf{D}_i$ may be chosen as a two-point upwind stencil

$$c \cdot \frac{1}{x_i - x_{i-1}} \cdot [-1, \quad 1, \quad 0] ,$$

a central difference stencil

$$c \cdot \frac{1}{x_{i+1} - x_{i-1}} \cdot [-1, \quad 0, \quad 1]$$

(convection term) and a three point centred Laplacian

$$a \cdot \frac{4}{(x_{i+1} - x_{i-1})^2} [1, \quad -2, \quad 1]$$

(diffusion term). Mixed derivatives like $\frac{\partial^2}{\partial x_0 \partial x_1} u$ can be approximated by the combination of an approximation of the uni-axial derivatives as

$$\frac{\partial^2}{\partial x_i \partial x_j} u = \frac{\partial}{\partial x_i} \circ \frac{\partial}{\partial x_j} u$$

$$\approx \mathbf{B}_{I\setminus\{i\}}^{-1} \circ \mathbf{D}_i \circ \mathbf{B}_{I\setminus\{i\}} \circ \mathbf{B}_{I\setminus\{j\}}^{-1} \circ \mathbf{D}_j \circ \mathbf{B}_{I\setminus\{j\}} u \tag{2.35}$$

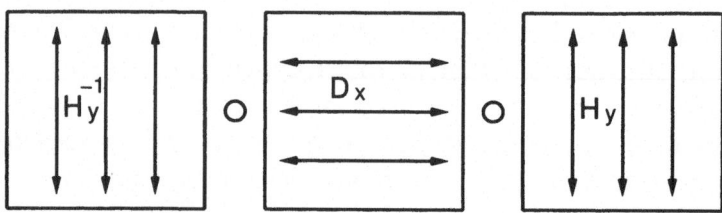

Figure 2.7. *Scheme for a finite difference operator in x-direction.*

or as the symmetrised version

$$\frac{\partial^2}{\partial x_i \partial x_j} u = \frac{1}{2}\Big(\frac{\partial}{\partial x_i} \circ \frac{\partial}{\partial x_j} + \frac{\partial}{\partial x_j} \circ \frac{\partial}{\partial x_i}\Big)u. \tag{2.36}$$

On adaptively refined meshes, the nearest neighbour nodes are chosen, which may lead to asymmetric stencils, i.e. non-uniform one-dimensional stencils.

$$a \cdot \frac{2}{(x_{i+1} - x_{i-1})(x_i - x_{i-1})(x_{i+1} - x_i)}[(x_{i+1} - x_i),\ \ (x_{i-1} - x_{i+1}),\ \ (x_i - x_{i-1})]$$

Further higher order modifications of the stencils have been tested as well. In the presence of a transport term in the equation, the unsymmetry is believed to be no problem, because the equation system is unsymmetric anyway. There are many ways to create discretisations of all kind of equations, e.g. for the Navier-Stokes equations by Schiekofer [266] or some hyperbolic conservation laws by Griebel and Zumbusch [153].

Following the usual convergence theory of finite difference schemes for elliptic PDEs, consistency and stability have to be shown. There is not that much theory known for sparse grid discretisations. However, the consistency error has been analysed for some model problems. It behaves like the consistency of regular grids, see Schiekofer [266] and Koster [190]. The second ingredient of a convergence analysis of the stability is still missing, but numerical experiments indicate that the stability deteriorates by a logarithmic factor, which results in similar convergence results as the Galerkin method on sparse grids.

The algorithmic part of the finite difference discretisation is much simpler than for the Galerkin method if one accepts the fact that the discretisation is unsymmetric. As usual only the matrix multiply is implemented and the matrix is not assembled for complexity reasons. The sparsity pattern, see Figure 2.8, shows that the matrix is not really sparse and a matrix assembly would deteriorate the overall complexity. Analytically it is known that the average

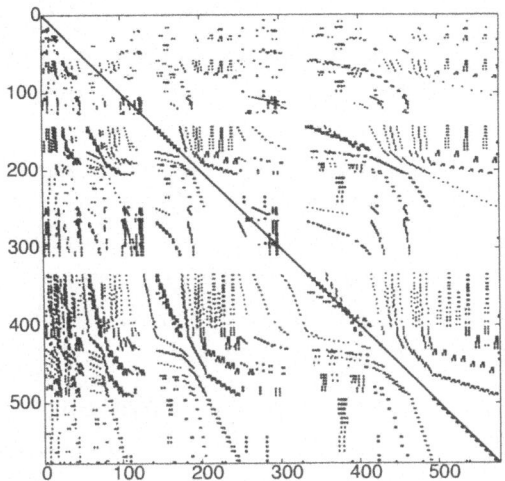

Figure 2.8. *Sparsity pattern of a two-dimensional finite difference discretisation.*

number of non-zero entries per row in the stiffness matrix with $N \approx 2^l \cdot l^{d-1}$ unknowns is roughly $2^l/l$ for the Galerkin method, i.e. the matrix is almost dense, see Balder [17]. The finite difference discretisation results in a lower average of l^{d-1} entries, which is still too much to be stored in a matrix. Due to unsymmetric finite difference discretisation, any iterative solver for the equation system has to deal with the unsymmetry and methods like BiCGstab and GMRes are quite popular. Iterative solvers can be accelerated by a diagonal scaling in hierarchical basis, see Schiekofer [266], or in an (interpolet-) wavelet basis, see Griebel and Koster [142]. Again, the methods work well on sparse grids, but have been analysed only for the discretisation on standard meshes.

Algorithmically, three steps have to be implemented for a finite difference stencil on sparse grids:

- transform a vector of nodal values to hierarchical basis: $\mathbf{H}_{I \setminus \{i\}}$

- apply a one-dimensional finite difference stencil along the coordinate axis x_i: $\mathbf{D}_i$

- transform the vector back to nodal basis $\mathbf{H}_{I \setminus \{i\}}^{-1}$

The basis transformation can be implemented as a sequence of one-dimensional transformations $\mathbf{H}_i$ and $\mathbf{H}_i^{-1}$, each along one coordinate axis x_i. The hierarchical basis can be substituted by any wavelet or pre-wavelet system. Furthermore, a similar procedure can be applied to vectors given in the transformed basis. This may save operations for the preconditioner, but is preferable anyway for dimensions $d > 2$ large enough in order to minimise the number of one-dimensional transformations.

An alternative approach to discretisations on sparse grids has been proposed by Hemker [172]. Cell centred finite volume discretisations are used in a combination technique. However, special care is needed, since the nodes at the cell centres do not coincide for different grid levels. Some modifications with piece-wise constant interpolation are needed. Furthermore, a multigrid method exists, again without rigorous theory.

As we saw in this chapter, there are various methods for solving the equation system which arise from the discretisation of partial differential equations. A complete theory of subspace correction methods was developed which covered both multigrid and multilevel methods and domain decomposition methods. The choice of the subspaces, either as a nested sequence of spaces or as a geometrical decomposition of the domain, leads to both methods which have superior computational complexity than ordinary direct and iterative solvers. Furthermore, the discretisation and iterative solvers on sparse grids were discussed. Sparse grids are comparable to other higher order discretisations in that they approximate a solution with fewer degrees of freedom, but each degree of freedom is more expensive than for standard discretisations. Hence we demonstrated different sparse grid discretisations, each with various possible iterative solvers.

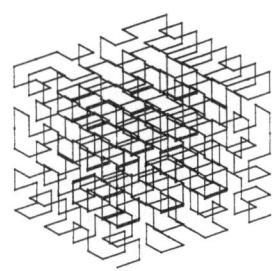

Chapter 3

Adaptively Refined Meshes

In the previous chapter the solution of linear equation systems was discussed. Often there are even optimal order methods, so that an equation system of n equations can be solved in $\mathcal{O}(n)$ operations. This is optimal in the sense that any method which produces an output of n numbers has at least $\mathcal{O}(n)$ complexity, even if the method is as simple as *write the vector zero*. Of course, one may be able to tune given methods or find new ones with a smaller constant in the complexity, but in general a lower limit is reached. Nevertheless, it may be desirable to solve a given PDE even faster. In this book we will present two ways to do this; we introduce parallel algorithms in chapter 5. This is a general approach to reducing the complexity by using more than just one processor. Another way to reduce the complexity is to change the discretisation and to use fewer degrees of freedom. This can be done for functions which are smooth or regular enough by a higher order or by a sparse grid method, see also chapter 2.5. In cases of non-smooth functions, adaptive mesh refinement can be used to concentrate the computational effort to regions where it is needed most. We will cover the topic of adaptive mesh refinement in this chapter.

First of all, we look at the discretisation of the PDE on a mesh. Both finite differences and finite elements are introduced. Although the schemes are fairly standard, it is interesting to see how they can be applied to adaptively refined meshes. Finite difference stencils with non-equidistant stencil lengths require special treatment in order to conserve the approximation order. Finite element discretisations can be applied to meshes with hanging nodes, if they are assembled and eliminated in the right way. In general, a symmetric stiffness matrix for a selfadjoint differential operator seems to be natural. However, boundary conditions and adaptively refined meshes can easily destroy this symmetry.

After this basic introduction of discretisation schemes, the main adaptive refinement cycle is discussed, which contains a posteriori error estimation and mesh manipulation. A brief overview over standard error estimators is given, some of them related to the subspace splittings of section 2.2. In addition to error estimators and error indicators, various strategies exist to select nodes, edges or elements for refinement. Some of them are based on rigorous theory, namely a proof of convergence of the adaptive cycle, and others are more heuristic. The mesh manipulation algorithms heavily depend on the type of mesh and additional restrictions on the mesh, such as triangular or hexahedral mesh, hanging nodes, closure rules and element aspect ratios.

We conclude the chapter with a discussion of data structures suitable for algorithms on adaptively refined meshes. Since there are different types of meshes, there will always be different representations of the meshes in a computer. However, whole program codes are based on a set of data structures which is necessary to enable optimal complexity implementations of some algorithms. Some structures which are actually in use in several packages are reviewed, were chosen for efficient memory usage, for sufficient algorithmic functionality and for ease of use. Especially for the goal of parallelisation of code it is extremely useful to have small and easy structures and algorithms, since the parallelisation adds a great deal of complexity itself. For these reasons, an extremely simple structure or interface to adaptively refined meshes is proposed which uses key-based addressing. Furthermore, an implementation of a key-based addressing scheme is discussed by the introduction of hash storage. Both seem to be novel elements in the area of partial differential equations, see [151, 171, 256, 266, 267], but are common and have proven successful in large areas of computer science.

3.1   The Galerkin Method, Finite Elements and Finite Differences

In this section we consider the discretisation of partial differential equations. Special emphasis is put on adaptively refined meshes and possible difficulties for the discretisation due to effects of adaptively refined meshes, such as non-uniform element sizes or edge lengths and hanging nodes. Furthermore, the treatment of boundary conditions is discussed, both for finite differences and for finite elements.

3.1.1 Finite Differences

A straightforward way to discretise partial differential equations is standard finite difference stencils. An elliptic equations like

$$-\Delta u = f \text{ in } \Omega \qquad (3.1)$$

translates to a one-dimensional difference stencil on an equidistant mesh

$$-\frac{u_{i-1} - 2u_i + u_{i+1}}{h^2} = f_i. \qquad (3.2)$$

The d-dimensional stencils can be constructed as sum of one-dimensional stencils along the coordinate directions for an equidistant axis-parallel and rectangular mesh, which results in $(2d+1)$-stencils. For example, the two-dimensional five-point stencil can be written as

$$-\frac{u_{i-1,j} + u_{i,j-1} + u_{i+1,j} + u_{i,j+1} - 4u_{i,j}}{h^2} = f_{i,j}$$

on a $\mathbb{Z}^2$ mesh. This discretisation is of consistency error two for $u \in C^{3,1}(\Omega)$ and is stable, which guarantees second order convergence for smooth solutions.

Adaptive mesh refinement leads to a non-uniform mesh spacing. A one-dimensional finite difference stencil may look like

$$-\frac{2}{x_{i+1} - x_{i-1}} \left(\frac{u_i - u_{i-1}}{x_i - x_{i-1}} - \frac{u_{i+1} - u_i}{x_{i+1} - x_i} \right) = f_i \qquad (3.3)$$

for mesh points x_i and values at the points u_i. The stencil coincides with Equation (3.2) for the equidistant case $h = x_{j+1} - x_j$. However, in general the consistency in the maximum norm is only of order one. The rate of convergence at each point is still second order for smooth solutions $u \in C^{3,1}(\Omega)$, see Esser and Niederdrenk [115], Samarskij [263], and Großmann and Roos [156].

Dirichlet boundary conditions define the value of u at the Dirichlet nodes $x_i \in \Gamma_D \subset \partial\Omega$ as $\gamma(x_i)$. The nodes are not degrees of freedom and can be eliminated from the equation system by the removal of the respective row in the equation system (3.2). Alternatively, the values x_i be left in the equation system, if the respective rows simply read as $u_i = \gamma_i$. For a symmetric matrix representation, the values x_i have to be eliminated, but for the application of difference stencils within an iterative solver it is sufficient to set the Dirichlet values u_i to γ_i.

Boundary conditions of Neumann-type can be approximated by a finite difference representation of the derivative as $(u_1 - u_0)/h$. This substitutes the row in the equation system (3.2). The lower order approximation of the boundary conditions does not hurt the asymptotic convergence rate. For a symmetric operator and a prescribed normal derivative ψ, the difference stencil can be written as

$$\frac{u_1 - u_0}{h^2} = \frac{1}{h}\psi$$

on the Neumann boundary $\Gamma_N \subset \partial\Omega$. More complicated boundary conditions such as Robin conditions can be discretised similarly.

Hanging nodes, which occur for adaptive mesh refinement in two and more dimensions, if mesh cells of different size are neighbours, can be treated in different ways. Since it is more complicated to use finite differences on arbitrary polygonal shaped cells, geometrical mesh closure algorithms are seldom used and hanging nodes cannot be avoided. The value u_i at such nodes can be defined by interpolation on the coarser neighbour. The interpolation condition is used to eliminate the degree of freedom x_i from the system matrix, or it substitutes the row in the equation system (3.2). In this case the resulting discrete operator is no longer symmetric. Alternatively, hanging nodes can be treated as degrees of freedom, where a modified difference stencil is used. This stencil is based on the available nodes of the neighbour cells, similar to Equation (3.3). Again, the discrete operator is no longer symmetric in general.

More generally shaped domains can be discretised with finite differences by alternative constructions of the stencils. Theoretically, finite difference stencils can be derived by a finite element quadrature scheme. Note that the scaling of the matrix entries and the construction of the right hand side f_i differs, even if the stencils may coincide. Furthermore, box-schemes or finite volume schemes define difference stencils between general polygonally shaped mesh cells.

The finite difference stencils on an axis parallel rectangular mesh can also be modified for more general shaped domains according to Shortley and Weller [273], see also Hackbusch [168]. The idea is to change stencils of nodes close to the boundary. The distance h of a leg of the stencil is shortened to the actual distance to the boundary, and the entries of the stencil are changed in accordance with Equation (3.3). This can be implemented nicely for Dirichlet boundary conditions, which is shown in Figure 4.33. However, for Neumann-type boundary conditions where nodes on the boundary are also degrees of freedom, some nodes may have to be split and shifted to different locations due to the existence of more than one neighbour node in the interior of the domain. Furthermore the modified distances influence the condition number of the discrete operator and the performance of iterative solvers.

The main advantage of the finite difference method is its simplicity for the regular mesh case. In the context of iterative solvers, the discrete operator need not be stored in a sparse matrix, but the operator can be applied to a vector on the fly. The coefficients are easy to determine. This is also true for simple boundary conditions. However, if a symmetric discrete operator is needed, symmetry is hard to maintain for adaptive meshes due to Equation (3.3), for hanging nodes due to the interpolation condition or modified stencils, and for modified stencils close to the boundary. Here, it seems easier to assemble the system matrix without these features completely and to modify it symmetrically in a second step. For non-selfadjoint operators the matrix is not symmetric anyway and finite difference stencils can be applied straightforwardly.

3.1.2 Finite Elements

To circumvent these difficulties of finite difference schemes for the symmetric discretisation of self-adjoint operators and for reasons of a more advance theoretical base, Galerkin methods can be used. The Galerkin method has been introduced in chapter 2.5.2 for the discretisation of partial differential equations on sparse grids. However, it is mainly used as a foundation of finite element methods, see Braess [54], Brenner and Scott [72], Ciarlet [88], and Renardy and Rogers [250]. The variational form of the differential equation is. *Find $u \in V$ such that*

$$a(u, v) = f(v) \qquad \forall v \in V \tag{3.4}$$

with a suitable function space V and boundary conditions. For second order elliptic differential equations on a suitably shaped domain Ω the space can be chosen as the Sobolev space $H^1(\Omega)$. The Galerkin method substitutes the space V by finite dimensional space, so that the discrete version of Equation (3.4) turns into a equation system for the solution u. The finite element method further restricts this approach and provides a construction of a set of basis functions, which actually defines a space V.

The domain Ω is partitioned into a number of disjoint elements τ_i,

$$\begin{aligned} \overline{\bigcup_i \tau_i} &= \overline{\Omega} \quad \text{and} \\ \tau_i \cap \tau_j &= \emptyset \quad \text{for } i \neq j \,. \end{aligned}$$

A small function space $V(\tau_i)$ is defined on each element τ_i, for example, linear functions on triangles, tetrahedra and simplices and tensor products of linear functions on rectangles, hexahedra and hypercubes. A basis $\{\Phi_{\tau_i,k}\}$ for each function space $V(\tau_i)$ is chosen so that functions in the global space $H^1(\Omega)$ can be constructed easily. For this purpose, conforming finite elements require the

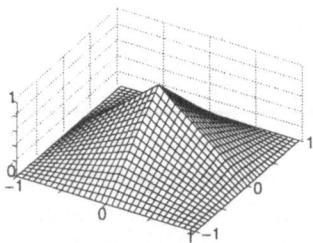

Figure 3.1. *A basis function defined on square shaped finite elements by bilinear local shape functions.*

functions to be continuous $C^1(\Omega)$. A straightforward way to ensure continuity is to use the Lagrange representation of $V(\tau_i)$ and to match coefficients on the boundary of an element $\partial \tau_i$.

The resulting functions ϕ_i in $H^1(\Omega)$ are piece-wise polynomial and have a local support, namely a limited number of elements τ_i. For linear functions or tensor products thereof, each function ϕ is related with one node of the mesh. The finite dimensional space V is defined as the span of the functions ϕ_i. The equation system is derived as a discrete version of the variational form $a(.,.)$

$$
\begin{aligned}
\sum_j a_{i,j} u_j &= f_i & \forall i \\
\text{with} \qquad a_{i,j} &= a(\phi_i, \phi_j) \\
\text{and} \qquad f_i &= f(\phi_i)
\end{aligned}
\tag{3.5}
$$

with coefficient vectors u_j and f_i. Each coefficient represents a single basis function, which is related to a single mesh node for Lagrange polynomials.

The stiffness-matrix A with entries $a_{i,j}$ is sparse with a limited number of non-zero entries per row, because of the local support of the functions ϕ_i. Its entries are computed by numerical quadrature or in simple cases analytically. For a second order scalar self-adjoint problem the bilinear form reads as

$$
a(u,v) = \int_\Omega \nabla u \, a \, \nabla v \, dx
$$

with a scalar coefficient function $a(x)$ or a symmetric tensor $a(x)$, both in $L_\infty(\Omega)$. The integral for the matrix entry $a(\phi_i, \phi_j)$ can be reduced to the domain $\text{supp}\phi_i \cap \text{supp}\phi_j$, which is at most a few elements τ. Furthermore, the integral can be partitioned into integrals over single elements τ_k, such that the functions $\phi_i|_{\tau_k}$ and $\phi_j|_{\tau_k}$ are polynomials. The quadrature on an element τ_k can be computed exactly for constant coefficients a and very accurate for smooth coefficients a by numerical quadrature formulae. The matrix assembly is usually organised in two steps. First the local matrices on a single element τ_k for the local basis functions in $V(\tau_k)$, also called finite element shape functions, are computed. In a second step the global matrix is assembled. Mathematically this means a summation of matrices according to the decomposition of the integration domain Ω into elements τ, but for an implementation this means a translation of local to global node numbers. While the local matrices are dense, the global matrix is sparsely populated. For higher order discretisations and for systems of partial differential equations or vector valued solutions, the global matrix is described more precisely as a sparse matrix of dense blocks.

Later on, boundary conditions can be applied to the matrix A and the right hand side f, e.g.

$$u \; = \; \gamma \; \text{ on } \; \Gamma_D \subset \partial\Omega \,.$$

Dirichlet boundary conditions prescribe the value of $u_i \in \Gamma_D$ at the Dirichlet nodes, due to the interpolation property of the functions ϕ_i. The degree of freedom is fixed for u and cannot be varied by a test vector of the weak formulation Equation (3.4). Hence row i of the equation system (3.5) has to be removed. In order to eliminate the value u_i completely from the system, the column i has to be eliminated, which leads to a modification of the right hand side $\hat{f}_j = f_j - a_{j,i}\gamma_i$ with the Dirichlet value γ_i. Moreover, the row i and the column i of the matrix A can be removed with $u_i = \gamma_i$. Alternatively, if the removal and possible re-numeration of the unknowns is too complicated, the equation system can be modified so that row and column i are the ith unit vector e_i, so that row i reads $u_i = \gamma_i$.

Neumann boundary conditions are treated differently. The natural derivative, which is the normal derivative for constant coefficient problems, is prescribed on the Neumann boundary Γ_N

$$(a \, \nabla u)n \; = \; \psi \; \text{ on } \; \Gamma_N \subset \partial\Omega \,.$$

Basically the right hand side is modified for a Neumann node i by the term

$$\hat{f}_i \; = \; f_i + \int\limits_{\Gamma_N} \psi\phi_i \, ds \,, \tag{3.6}$$

which vanishes for zero prescribed derivatives. Mixed boundary conditions and conditions on derivatives other than Neumann conditions can be written in the more general form of Robin boundary conditions

$$(a \, \nabla u)n + \sigma u \; = \; \psi \; \text{ on } \; \Gamma_R \subset \partial\Omega \,.$$

Here, the stiffness-matrix has to be modified by the term

$$\hat{a}_{i,j} \;=\; a_{i,j} + \int\limits_{\Gamma_R} \phi_i \sigma \phi_j \; ds$$

in addition to the right hand side similar to Equation (3.6).

Geometrically non-conforming meshes have elements τ_i, which do not nec-
essarily share complete edges or faces $\overline{\tau}_i \cap \overline{\tau}_j$. The construction of continuous
functions $\phi_i \in C^1(\Omega)$ is not as simple as it is for conforming meshes, where
coefficient matching is sufficient. Generally, some linear interpolation conditions can be defined, which guarantee global continuity. The local stiffness
matrices are assembled as usual, but the global assembly step includes the
interpolation conditions. This can be annoying, since the interpretation of the
degrees of freedom as values of the discrete solution at the nodes of the mesh
may fail. Furthermore, it can be computationally expensive to incorporate
the interpolation conditions, especially if the global stiffness matrix has to be
symmetric.

In an adaptive mesh refinement procedure, a special type of geometrically
non-conforming mesh can occur: an element τ_0 is sub-divided into several
smaller elements $\tau_{0,i}$, while a neighbour element τ_1 remains the same. In this
special case the interpolation condition for nodes x_j on the common boundary
$\Gamma = \overline{\tau}_0 \cap \overline{\tau}_1$ can be formulated as

$$u|_{\tau_{0,i}}(x_j) \;=\; u|_{\tau_1}(x_j) \,,$$

where nodal values on one of the elements are on the left side and linear
combinations of nodal values of τ_1 are on the right hand side. Hence this can
be used as a definition of the nodal values in the interior of Γ. The nodes are
called hanging nodes . They are not degrees of freedom in the global stiffness
matrix, but are eliminated in the global matrix assembly step. Alternatively,
they can be eliminated in a post-processing step. In the finite element context,
this can be rewritten as

$$u_i \;=\; \sum_k u_k \Phi_{\tau_1,k}(x_j)$$

and interpreted as an exchange of shape functions of the refined elements
$\Phi_{\tau_{0,l;k}}$ by the shape functions of the parent element $\Phi_{\tau_0;k}$ for the conforming
situation at node x_j. The global stiffness matrix can be modified by elimination
accordingly or by assembly with the appropriate shape functions.

Other types of variations of the described finite element scheme include
isoparametric elements, where a conformal mapping is used between a polygonal shaped element τ_i and its final shape with curved boundary on Ω. This

mapping has to be taken into account for the numerical quadrature of the local stiffness matrix. Different non-interpolating or non-local shape functions can be used for spectral methods, which are usually based on Fourier expansions and FFT, pseudo-spectral and higher order methods, where also energy minimising shape functions are preferred, see Karniadakis and Sherwin [182] and Zumbusch [322]. Other modifications include Petrov-Galerkin methods for non-selfadjoint operators, where the space for the solution u and test functions v are chosen differently to stabilise the discretisation of convection and transport terms and non-conforming finite elements, where the global functions ϕ_i are not continuous and the local shape functions are coupled only weakly.

3.2   Error Estimation and Adaptive Mesh Refinement

We saw how to discretise a PDE on an adaptively refined mesh. However, we need to discuss how such meshes are created. We describe a cycle of adaptive mesh refinement, where a solution of the PDE on a given mesh serves as a starting point for a sequence of refined meshes. Since we assume that we do not have enough a priori knowledge to choose a good mesh, we need some procedure to improve the given mesh. This can be done by an error indicator or an a posteriori error estimator, which indicate regions of the mesh where higher resolution is needed. In a second step the mesh has to be refined accordingly. This procedure is based on the assumption that a local refinement of the mesh actually reduces the approximation error in that region, which is clearly the case for an elliptic operator and local discretisation schemes like finite elements or finite differences. Furthermore, the assumption is needed that an equilibrated error does indeed indicate an optimal mesh, which basically is a sort of Pareto optimality, see Babuška and Rheinboldt [14]. In the case of wavelet schemes, such an assumption is fulfilled for best n-term approximation, see DeVore and Popov [102].

The discretisation error ϵ is the difference of the exact solution of the PDE and the solution obtained by the discretisation, if the equation system is solved exactly. It depends on the mesh, on the solution of the PDE, on the discretisation scheme and the norm the error is measured in. The convergence rate of a discretisation scheme is the behaviour of the discretisation error as the mesh size $h \to 0$ goes to zero and the mesh becomes finer and finer. Let us assume that we use a second order discretisation with an error decay of $\mathcal{O}(h)$ in the energy norm. This is usually true for sufficiently regular solutions only, e.g. $H^2(\Omega)$. However, in the case of singularities, the observed convergence can be much slower, i.e. $\mathcal{O}(h^\alpha)$ with $0 < \alpha < 1$ due to the presence of strong

| discretise PDE on current mesh |
| solve equation system |
| estimate local errors η_i |
| flag nodes for refinement |
| create new mesh and |
| enforce geometric conditions |
| until global error $\tilde{\epsilon} \leq$ tol |

Figure 3.2. *The adaptive mesh refinement algorithm.*

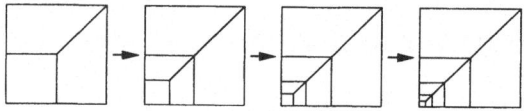

Figure 3.3. *A sequence of a-priori refined conforming isoparametric quadri-lateral meshes.*

stress concentrations in the vicinity of crack tips and obtuse corners, boundary layers, changes in the coefficient functions and other effects. Then, instead of $n = tol^{-d}$ unknowns we have to use at least $n = tol^{-d/\alpha}$ unknowns on a uniform mesh for a given error tolerance *tol*.

One way to overcome the problem of slow convergence due to singularities is to employ adaptive mesh refinement, see Figure 3.3. The idea is to use small h in the vicinity of singularities and stress concentrations only. An analysis of the local amount of work and the local discretisation error reveals that an equidistribution of the local error gives a good strategy for mesh refinement. In the ideal case, the performance of a discretisation on adaptively refined meshes in the presence of singularities is comparable to the performance of the original uniform mesh discretisation for regular problems with respect to discretisation error versus number of degrees of freedom. Such meshes can be constructed by best-n-term approximation of the solution.

A simple way to use adaptive meshes is to use information available before the computation (a priori) and to refine the mesh appropriately. This can be incorporated into the mesh generation process, where estimates of error densities or local mesh sizes h are prescribed. The optimal mesh refinement depends on the solution, the error norm and the prescribed error tolerance. Hence, a priori information is often not sufficient. One way to improve the mesh refinement process is to construct a feedback loop: the PDE is solved several times on different meshes. The actual solution is used to guide the mesh refinement for the next solution. This can be done either with a complete re-meshing and numerical error density functions or by refining the actual mesh by local modifications. The adaptive refinement itself is guided by an error indicator (a posteriori), which flags regions of large local errors. For early implementations see Rheinboldt and Mesztenyi [253], Mesztenyi, Miller and Szymczak [211], and Kelly, Gago, Zienkiewicz and Babuška [184].

The simplest way to adapt the mesh to the solution is to use a priori information in a one-step process. First a mesh is created and then the equation system is solved. The a priori knowledge, where special features of the solution such as singularities are likely to occur can be used to create a graded mesh, which has small mesh sizes h next to these features. The mesh refinement can be controlled by hand or with an error density function, which defines an estimate of the error and the need for mesh refinement on the domain Ω. The error density can be fed into the mesh generation process.

A feedback cycle around this process leads us to an adaptive mesh refinement process. Now the error density function can be based on the computed solution of the previous cycle, which substitutes the a priori information. The local errors η_i are estimated on the previous mesh and interpolated onto the domain Ω. This gives the error density, which again is fed into the mesh generation process. Usually a complete re-meshing is used. This process fits nicely into an existing frame of mesh generation and FEM software.

3.2.1 Error Indicators and Estimators

Alternatively to heuristic error density functions, more precise a posteriori error estimators can be used. In a feedback approach, the PDE is solved several times on a sequence of meshes. Here, the most recent solution is used to guide the construction of the next mesh, where an a posteriori error estimator $\eta(x)$ estimates the discretisation error $\epsilon(x)$ of the most recent solution. The discrete version of the error estimator η_i is related to either elements, nodes, edges, or faces of the mesh. It should be connected to the local errors ϵ_i by

$$c \cdot \epsilon_i \leq \eta_i \leq C \cdot \epsilon_i. \tag{3.7}$$

An analysis of the local amount of work and the local discretisation error reveals that the optimal mesh has equidistributed local errors ϵ_i. Furthermore, the optimal mesh consists of $n = C_\alpha \cdot \text{tol}^{-d}$ unknowns, which is equivalent to a convergence rate independent of α, see Babuška and Rheinboldt [14]. Hence, the mesh is refined at locations with large estimated error η_i. This can be interpreted as an optimisation process to achieve an equidistribution of the η_i. Equation (3.7) guarantees the efficiency of the process, see Russell and Christiansen [260]. Neither are regions with large errors missed, nor is unnecessary refinement performed. Although Equation (3.7) can only be proved under restrictive assumptions (sufficient regularity, saturation, etc.), asymptotically for a mesh size $h \to 0$ small enough, such bounds can be showed for many popular error estimators, see [2, 15, 179, 249, 299].

However, the main advantage of an error estimator compared to an error indicator is that the local η_i give an estimate for the global error ϵ

$$\epsilon \approx \sum_i \eta_i ,$$

which can be used for a termination criterion of the adaptive refinement cycle, see Figure 3.2. In the absence of error estimators, error indicators can trigger local mesh refinement and the termination criterion has to be based on other data, such as the energy norm of the solution compared to previous results. However, computations also have to terminate when they run out of computer resources such as memory or computing time available. Hence it is not possible to prescribe a tolerance in advance and compute until the estimated error is small enough, but the computation terminates when the resources are exhausted and the estimated error is a result.

Construction of early error indicators and error estimators was purely heuristic. Here simple local gradients were usually employed. In Bank and Weiser [22] and in Babuška and Miller [13] local problems were set up and solved. This approach led to residual based, Dirichlet based and Neumann based local error indicators. Alternatively local sub-problems with higher order discretisation were also used to construct error indicators by Craig, Zhu and Zienkiewicz [93]. A modern, rigorous theory for the construction of local and global error estimators, which is based on the dual problem approach, can be found in Becker, Johnson and Rannacher [30], and Becker and Rannacher [31], which has traces back to dual error estimates by Oden, Demkowicz, Rachowicz and Westermann [225] and Sander and de Veubeke [265]. The norm equivalence (2.11) can also be used to derive error estimators, for details see Oswald [231, chap. 5], [230]. Note finally that for simple elliptic operators it is possible to get rid of the so-called saturation assumption, which is usually a prerequisite in the construction and analysis of error estimators, see e.g. Dörfler [105], Dahlke, Dahmen, Hochmuth and Schneider [96], Dahmen [97], and Cohen, Dahmen and DeVore [91].

Such methods also include strategies to mark elements for refinement. Heuristics have been used to choose elements with ϵ_i above some threshold, which again is calculated by some fraction of the maximum or mean error or some extrapolation, see Babuška and Rheinboldt [14] and Russell and Christiansen [260], or a certain fraction of the elements or of the global error ϵ.

3.2.2 Mesh Refinement

The mesh can be refined by a complete new *re-meshing* of the domain Ω guided by the error density η_i, by *overlaying patches* of finer mesh size h over regions of refinement, or by local *element-wise* mesh refinement, where some elements are substituted by other, smaller elements.

The re-meshing process can be quite expensive, because mesh generation is expensive for complicated geometries Ω in general. Furthermore, local details of the solution and the error, which are information on the previous mesh, is lost due to interpolation to the new mesh. This can cause trouble for the local mesh refinement, because the local error estimate heavily depends on the underlying mesh. A multigrid algorithm for the solution of the equation system can either be based on the non-nested sequence of meshes or it can be an algebraic multigrid applied to the present mesh.

A more popular way to construct a feedback cycle is based on nested meshes. The previous function space is a subset of the function space on the new mesh. The mesh is obtained by some modifications of the previous mesh. The nesting of the spaces gives an improved interpolation of the solution and the error estimates. Multigrid methods can exploit the nested sequence of space directly. One way to modify the mesh is to define rectangular patches, where refinement is required. A finer mesh of a mesh size of say $h/2$ is overlaid on the previous mesh. Iterating this gives a sequence of patches of different mesh sizes and area, resolving local features of the solution on different scales, see Berger and Oliger [35], and Berger and Colella [34]. The finest mesh composed of the patches is the union of all function spaces of the coarse mesh and the patches. This can also be used as a nested sequence of spaces for a multigrid solver. However, specialised multigrid methods have also been developed for this composite type of mesh, see Brandt [64], Hackbusch [166], McCormick [209], and Figure 3.4 (left).

A similar mesh refinement technique for more general geometries and meshes is based on local mesh refinement. A given mesh is processed element by element. An element will be substituted by several other smaller elements if local refinement is required. A quadrilateral can be bisected and substituted by two quadrilaterals of half the area. Alternatively, it can be replaced by four similar quadrilaterals of one fourth the area, see Rheinboldt and Mesztenyi [253] and

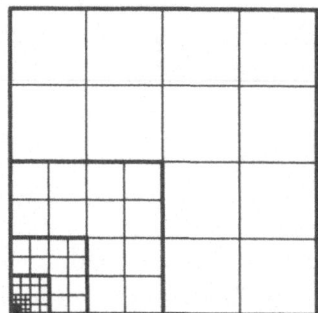 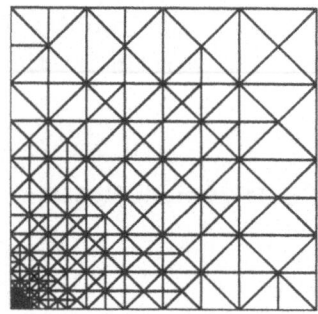

Figure 3.4. *Adaptively refined meshes. Patch-wise refinement (left) and element-wise refinement (right).*

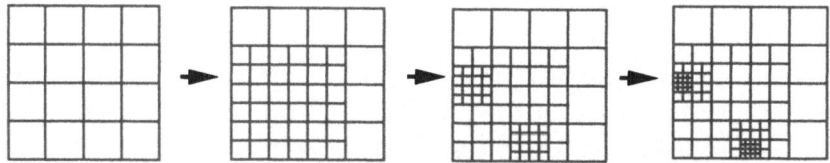

Figure 3.5. *A sequence of adaptively refined non-conforming quadrilateral meshes.*

Figure 3.5. Analogously hexahedra can be bisected or cut into eighth parts. A bisection strategy should take care that the elements decrease in size along all coordinate directions in order to control the maximal diameter of the elements $h_{\max}$ for convergence by changing the direction of the bisection. A subdivision of an element into 2^d elements does not affect the aspect ratio of the elements and gives local $h_{\max} \to 0$ anyway. The resulting meshes contain hanging nodes where elements of different size are joined. This always happens when the mesh is refined locally. Hanging nodes represent a constraint for a conforming discretisation, which can be eliminated from the equation system. They are not degrees of freedom.

Next we consider triangular and tetrahedral elements. There are algorithms which provide element-wise refinement without hanging nodes. Meshes based on triangular and tetrahedral elements can be constructed for arbitrary polygonally shaped domains and allow for easier representations of complicated geometries Ω. The idea for mesh refinement is to bisect elements or to subdivide them into 2^d elements as above, see Figures 3.7 and 3.4 (right). In

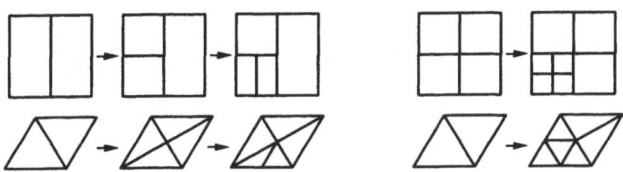

Figure 3.6. *Element-wise mesh refinement with quadrilateral and triangular elements. Bisection (left) and subdivision into four elements (right).*

addition to this refinement, which is triggered by an error estimate, further refinement is used to eliminate hanging nodes and to smooth the refined mesh. This can be done along some geometric rules, see Figure 3.6. A major concern here is to maintain angle conditions of the mesh for reasons of the discretisation error. Here, a bounded minimal interior angle ϕ

$$0 < c < \phi < C < \pi$$

or at least a bounded maximum interior angle independent of the mesh size h is desirable. For the interpolation properties of the FEM, however, the maximum angle condition is sufficient. The minimum angle condition is necessary for the performance of iterative solvers. If one bisects edges opposite to one node too often, the minimum angle condition will be violated. Several bisection strategies such as *longest edge* and *newest node* have been developed so far which choose the edge for bisection in a way to guarantee such bounds, see Rivara [255] and Mitchell [213]. The subdivision of an element into 2^d elements, *red refinement*, also requires some bisection, *green refinement*, although the subdivision of a triangle itself does not introduce any dangerous angles. This is due to the removal of hanging nodes, which triggers a sequence of element bisections, see Bank, Sherman and Weiser [21]. The subdivision of tetrahedral elements into 2^d elements requires a more complicated stable subdivision scheme, because at least four of the resulting tetrahedra cannot be similar to the original tetrahedron, see Bey [40] and Bey [41] in the general case.

Mesh refinement algorithms differ from numerical algorithms described so far as they operate on graphs. Local refinement can be implemented very simply, but closure rules or a 1 : 2 irregular grid condition impose further restrictions on the mesh. Efficient algorithms to enforce these conditions are also $\mathcal{O}(n)$ algorithms, where n however denotes the number of elements of the final mesh instead of the initial mesh. The problem is that some initial mesh refinement may trigger a sequence of further refinement, which is difficult to

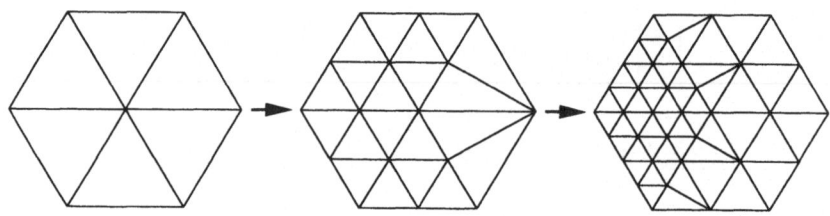

Figure 3.7. *A sequence of adaptively refined conforming triangle meshes.*

estimate. However, using a stack of elements that need to be checked for the rules, each modification of a mesh only causes a finite number of neighbour or parent elements to be checked. Finally, the amount of work is proportional to the number of mesh modifications, which is related to the number of new elements. This is true for bisection algorithms by Rivara [255], for red-green refinement, see Leinen [197], and for 1 : 2 irregular meshes, see Griebel and Zumbusch [152]. Note that also the parallelisation of such algorithms differs from the parallelisation of numerical algorithms and that the number of communication operations cannot be bounded a priori. This also lead to modifications of the mesh refinement rules at the processor partition boundaries, see Bastian [26].

Finally, there are many more variations of mesh refinement algorithms and rules. Rules exist for the refinement of mixed meshes consisting of several element types such as triangles and quadrangles in two dimensions and tetrahedra, hexahedra, prisms and pyramids in three dimensions. Furthermore, isoparametric distorted quadrangles can be used to eliminate hanging nodes and create a closure of a mesh containing refined quadrangles.

A more recent generalisation of the adaptive element refinement process introduces anisotropic refinement. A priori estimates usually require isotropic refinement, $h_{\max} \to 0$. However, near the boundary, next to rapid changes in the coefficients (material) or inside of thin structures or within anisotropic materials, anisotropic behaviour of the solution can be observed. An optimal mesh for such a situation would use small h in directions of rapid changes of the solution and larger h in normal directions, like sparse grids at the boundary or *blue* refinement by Kornhuber and Roitzsch [188]. The algorithms discussed so far cannot construct an anisotropic mesh. Modifications to include anisotropic refinement include local error estimates which depend on the direction and preferably element bisection along the directions of interest.

3.3 Data Structures for Adaptively Refined Meshes

The adaptive mesh refinement algorithms discussed so far need to be implemented carefully to ensure an acceptable computational complexity. While the implementation of local error estimators can be very similar to the discretisation of the PDE or a single solution step, care has to be taken that the amount of work for error estimation and mesh refinement does not dominate the actual solution. This can often be achieved easily. However, for the estimators based on the solution of a dual problem this should be ensured. Nevertheless, the pure mesh manipulation part of the mesh refinement might also be quite expensive when it comes to closure algorithms to remove hanging nodes or algorithms to fulfil other restrictions of the mesh like element aspect ratios.

Some of the operations are purely local and involve neighbour elements or neighbour nodes only. Hence, an appropriate data structure of an implementation should be able perform these operations in a bounded, local amount of time. In order to discuss some possibilities of a data layout, we will cover different ways to represent the geometric mesh and the discretisation of the PDE.

3.3.1 Structured Meshes

We start with the representation of structured meshes. We consider a rectilinear domain

$$\Omega = [a_1, b_1] \times [a_2, b_2] \times \ldots \times [a_d, b_d] \ ,$$

which is discretised on a rectilinear mesh

$$g = [1, \ldots, n_0] \times [1, \ldots, n_1] \times \ldots \times [1, \ldots, n_d] \subset \mathbb{N}^d \ .$$

The nodes of the mesh are located at g, while an element is located between 2^d nodes. A d-dimensional array can be used to store functions on the mesh efficiently. Usually the array will be mapped to a one-dimensional contiguous chunk of memory. The data structure is as simple as an array available in high level programming languages, where slight implementation differences exist in the order of the dimensions. The mesh can be generalised to curvilinear meshes, where an additional mapping of nodes is used to approximate more general, curved domains Ω. Furthermore, the mesh spacing, i.e. the mapping of g onto Ω, may be irregular and concentrate nodes in certain areas of the domain.

A discretisation of finite difference type can be set up immediately on a structured mesh. The differential operator can be applied to a function on the mesh by a loop over all nodes x and the retrieval of its neighbours (e.g. $x \pm he_i$). Furthermore, finite element discretisation can be used on structured meshes quite easily. For a constant operator, a finite difference stencil can be pre-computed and applied which is equivalent to the appropriate finite element discretisation. In general, a stiffness matrix needs to be assembled by the finite element method. However, due to the structure of the mesh, the pattern of the non-zero entries of the matrix is known in advance. It is a structured pattern of sub- and super-diagonals, so that a banded matrix storage scheme can be used. Other matrix schemes may be appropriate for direct equation solvers. However, both structure of the mesh and of the matrix are known in advance. Basically the numerical data can be stored in a structured fashion and very little additional information needs to be stored. The resulting data structures are straightforward.

3.3.2   Unstructured Meshes

Unstructured meshes can be used to describe more general domains. One or more element types are connected to tesselate the domain Ω. If we consider linear finite elements, so that the degrees of freedom are located at the nodes, the coordinates of the nodes have to be stored. Furthermore, the nodes have to be accessible in an implementation and operations like a loop over all nodes should be possible. Some connectivity between nodes is needed for a discretisation on an unstructured mesh. Hence it is useful to store elements, since the number of nodes per element is known in advance, while the number of adjacent nodes or elements of a single node can only be bounded by angle conditions. Furthermore, edges may be useful as a data structure: the sparse matrix representation of the stiffness matrix can be stored in the edge data structures, where a matrix entry $a_{i,j}$ between two adjacent nodes i and j corresponds to the (directed) edge. Faces in three dimensions and edges are also necessary for higher order methods, for non-continuous methods and for the implementation of certain types of boundary conditions.

The number of elements can be stored in a single vector of indexes, where a sequence of indexes represents the nodes of an element. In the simplest case, the indexes point into a vector of coordinates of the nodes and into some data vectors. A similar functionality can be achieved with a linked list of elements, where each elements consists of pointers to nodes. Data can either be stored in the node data structure, which is also organised in a linked list, or an index in the node points into data vectors. The basic operations on these data structures are: loop over all elements, find the nodes of an element, and loop over all nodes. The loop can operate on a linked list, which would be easier

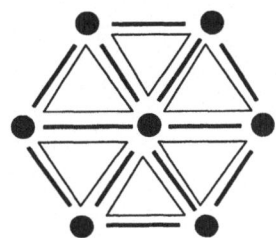

Figure 3.8. *Representation of a two-dimensional unstructured mesh as elements, nodes and edges.*

to modify and to extend, or on a vector of entities. It is of course possible to introduce edge or face data structures or additional pointers from one entity to another one. This mainly depends on whether the link is necessary in a certain stage of an algorithm or can be reproduced during the algorithm quickly. Often it is a trade-off between memory consumption and run-time to compute the necessary connection. However, some pointers or links are necessary in order to avoid search operation which would spill the overall computational complexity, due to e.g. a linear search over all elements.

3.3.3 Sparse Matrices

A finite difference discretisation on an unstructured mesh usually needs to be assembled into a matrix, since the local element matrices differ due to the different shapes of the elements. The storage of local stiffness matrices would require more memory than a globally assembled matrix. Since there is no single standard enumeration of the nodes, the sparsity pattern of the stiffness matrix is not regular and is unknown in advance. Instead of band matrices, general sparse matrices have to be employed. Many standard sparse matrix storage schemes like Yale Sparse Matrix Package, ITPACK and others are comparable to the following row-indexed sparse matrix scheme, see also Figure 3.9.

One single vector v contains all non-zero entries $a_{i,j}$ of the matrix. The entries are ordered row-wise and perhaps also within each row. A vector c of the same size contains the column number j of each of the entries, so that c_k contains the column index of the entry v_k. A third vector r of the size of the number of rows contains the index of v of the first entry in a row. Hence all entries $a_{i,j}$ are stored in the locations between v_{r_i} and $(v_{r_{i+1}} - 1)$. The respective columns can be found at the same locations in vector c between c_{r_i} and $(c_{r_{i+1}} - 1)$. A matrix multiplication can be implemented with this data

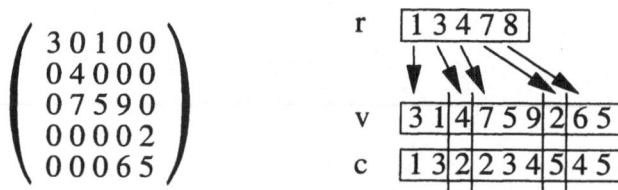

Figure 3.9. *A row-indexed sparse matrix storage scheme.*

structure in a complexity linear in the number of non-zeros, with a number of write operations as the number of rows. Furthermore, the transposed matrix multiply is of the same complexity, however with write operations proportional to the number of non-zeros, which can be computationally expensive. The storage scheme for usual finite element discretisations with several entries per row leads to an average memory consumption of one floating point and one integer value per non-zero entry. Although it is possible to further reduce the amount of memory and to eliminate the integer vectors if information available from the unstructured mesh is used instead, the sparse matrix storage is fairly compact. Further compression is possible for symmetric matrices, where only the upper triangle is stored. However, the number of write operations per matrix multiplication increases as for the transposed matrix multiplication, since part of the matrix is stored row-wise and the remaining part column-wise. The matrix storage scheme is independent of any geometric or other data structure and can be used in direct and iterative solvers.

It is difficult to extend the sparse matrix storage in the present form due to fill in of a direct solver, the introduction of boundary conditions or generally a change in the number of degrees of freedom. However, some extensions of this basic matrix storage exist which provide such possibilities at the expense of slightly larger memory consumption.

There is another reason to use data structures like vectors, matrices and meshes. In an adaptive mesh refinement environment the total number of degrees of freedom, the number of elements in a mesh and the number of matrix entries changes during a computation due to mesh manipulations. Thus some implementations only provide a dynamically changing mesh data structure and put all numerical data into this mesh. Hence all numerical algorithms operate on the mesh. However, it seems to be much cleaner to separate numerical algorithms of linear algebra, discretisation algorithms like a finite element assembly procedure and mesh manipulation algorithms. The entities of a vector, matrix and a geometrical mesh provide exactly the abstractions needed for the appropriate data containers for this separation. Usually this also leads to more efficient implementations, since single parts of the code are separate, operate on smaller and less fragmented pieces of memory and can sometimes even be reused from libraries or other implementations. In this sense, the mathematical abstractions are very useful also for highly sophisticated adaptive implementations.

3.3.4 Adaptively Refined Meshes

Adaptive mesh refinement procedures change the computational mesh. The appropriate storage scheme depends on the type of mesh refinement. Based on structured meshes, patch-wise mesh refinement can be applied. The final mesh g_{fine} is the virtual superposition of a set of meshes g_i, each one of simpler structure and covering at least part of the domain Ω. We can think of structured meshes g_i with rectilinear or curvilinear shape. The computation takes place on the structured meshes g_i. Furthermore, interpolation and projection procedures exist between overlapping parts of meshes g_i and g_{fine}. Since efficient implementations of many algorithms on structured meshes and simple data structures for the meshes themselves and operators on the meshes exist, they can be used for patch-wise refined meshes as well. In a finite difference context, only additional interpolation operators have to be provided. Usually there are only a few structured meshes g_i, compared to a large overall number of degrees of freedom. Hence, some simple list of structured meshes, together with some geometrical description of the correlation and overlap, is sufficient to represent a patch-wise refined mesh. Note that this is also true for some simple types of parallelisation, be it as a distribution of mesh g_i or by a parallelisation of procedures on each mesh g_i, preferably by rectangular slices of the mesh.

Element-wise mesh refinement changes the computational mesh on a much finer scale. Hence a description of the refined mesh by patches does not seem to be appropriate. However, unstructured meshes can be used. Given an initial coarse mesh as a set of polygonally shaped elements, refinement rules transform the mesh into another set of elements. Since unstructured meshes can be stored as a list of elements and nodes, adaptively refined meshes can also be stored this way. An element of the mesh may be replaced by a set of smaller elements, and new nodes may be created and added to the list of nodes. Hence a fixed size vector of elements and nodes can be used for a single unstructured mesh, but a variable size vector or list is useful during adaptive mesh refinement. The differential operator can be assembled into a single sparse matrix. However, during mesh refinement, both the number of degrees of freedom and the entries

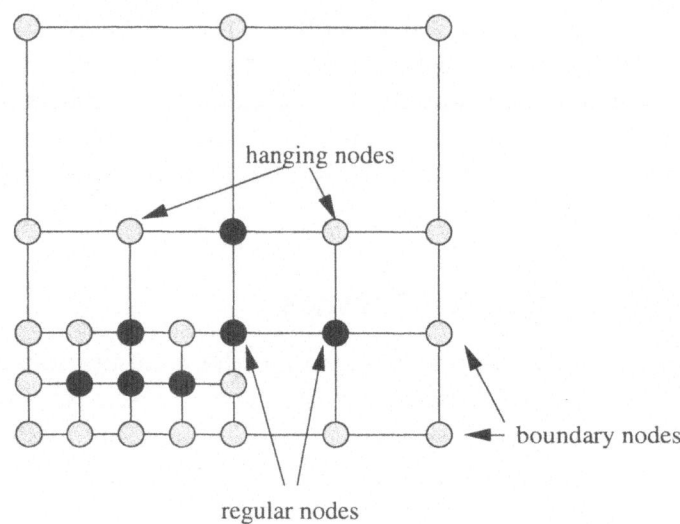

Figure 3.10. *Node types of an adaptively refined mesh of an initially structured mesh.*

of the stiffness matrix itself change. In the case of small changes in the mesh, it can be more efficient to use local stiffness matrices instead of one globally assembled matrix: the local matrices of elements, which are not changed, can be reused, while matrices of new elements have to be computed. However, local stiffness matrices require a larger amount of memory and slightly more operations during a matrix multiplication.

3.3.5 Key-Based Addressing

In the case of adaptive mesh refinement of an initial regular mesh, coordinate arithmetic can also be used to locate elements and nodes. Usually, indexes or address pointers are used to reference geometric objects. However, on a regular mesh, a unique and precise integer representation of the coordinates is also available. Adaptive mesh refinement, e.g. by bisection, also allows for an integer representation of the coordinate tuples, see Figure 3.10. Hence the coordinate values or some values derived by the coordinates can be used to describe a geometric object, see Figure 3.11 and [90, 256, 266, 285].

Given a coordinate tuple of a node and the local mesh size h on a rectangular mesh, we can compute the coordinates of neighbour nodes. For a given element described by the coordinates of its nodes, we can determine the coordinates of the vertices of its neighbour elements in a locally regular rect-

0																-1
0								1								-1
0				2				1				3				-1
0		4		2		5		1		6		3		7		-1
0	8	4	9	2	10	5	11	1	12	6	13	3	14	7	15	-1

Figure 3.11. *Key values of the nodes in five levels of a sequence of one-dimensional, refined meshes.*

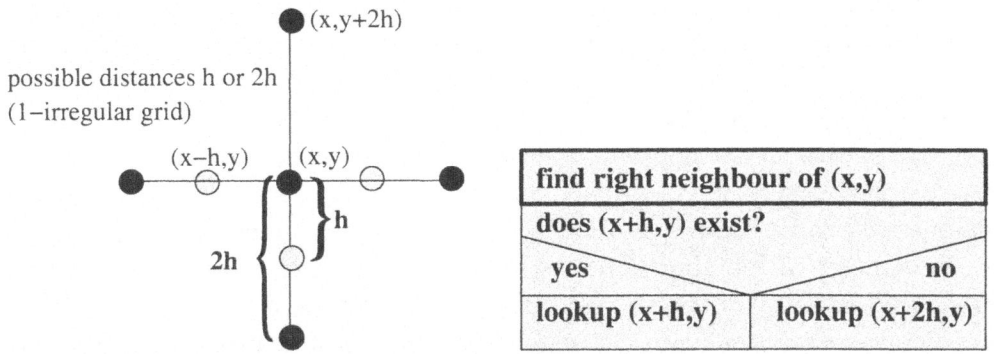

Figure 3.12. *Finite difference stencil on a one-irregular adaptively refined mesh and associated node look-up procedure.*

angular mesh. In order to use this scheme, we need a container data structure of nodes and elements where they can be retrieved by their coordinates. This can be done with various data structures, including sorted lists, search-tree, hash-tables and other basic structures used especially in the design of database systems, see Samet [264]. However, the computational complexity of the storage and retrieval is important. In particular, a linear access time is needed, compared to logarithmic access for many sorted structures. Hash-tables can provide linear access time characteristics in the statistical average, given a suitable mapping- or hash-function.

While address arithmetic is simple on regular uniform and rectangular meshes, the calculation of coordinates on adaptively refined meshes is not immediately clear. However, the mesh refinement procedure creates refined meshes with some properties that can be exploited for the calculation of nodes. A bisection scheme always gives local mesh sizes of power of one half $h = 2^{-i}$. The distance between one node and its neighbour node will be $h = 2^{-i}$. The coordinates of the node allow one to determine on which level l of mesh refinement the node is located, i.e. the minimum l so that the coordinates are a multiple of 2^{-l}. Now we can search for its neighbour along one coordinate direction by checking nodes at different distances $i = l, l+1, \ldots$. The data structure for the set of nodes allows one to query whether a certain node does exist. Given a distance 2^{-i}, $i \geq l$ with a neighbour, so that the node at distance $2^{-(i+1)}$ does not exist, we have found the nearest neighbour of the node. This property holds because the adaptive refinement procedure creates only direct children nodes, and as a consequence, with a single node its parent node is also present.

Further properties of mesh nodes are provided by rules on hanging nodes. If the number of hanging nodes along an edge is limited, the number of coordinate locations to search will also be limited. Given a mesh with square shaped elements and one-irregular meshes and, furthermore, the minimum mesh size $h \leq 2^{-l}$ at a node, neighbour nodes are located either at distance h or at distance $2h$ along each coordinate direction, see Figure 3.12. The mesh one-irregularity can be relaxed so that elements of any size meet at a node as long as the number of hanging nodes is limited by one per edge. Now the local mesh size may vary between h and $2^d h$, which leads to a maximum of $(d+1)$ coordinate query operations per neighbour. In a similar fashion, the elements can be identified based on the set of nodes only, which can be used for a finite element discretisation. Furthermore, control cells can be computed from the nodes for a finite volume scheme.

Similar key based addressing schemes can be obtained for other mesh refinement procedures and for different domains. For example, sparse grid nodes can be organised this way, see also Schiekofer [266]. Along a coordinate axis, one-dimensional trees of nodes are present. Instead of links or pointers to form the trees, a key based addressing scheme can be used to access parent and children elements along each coordinate axis. In this case, d parent and $2d$ children pointers can be eliminated by keys. The encoding scheme of Figure 3.11 has been used for sparse grids by Zumbusch [323].

Furthermore, key-based addressing can be established on unstructured meshes: the elements of a general triangulation or tetrahedrisation τ_0 of a polygonal domain Ω can be enumerated, e.g., by a space-filling curve. Along with a numbering scheme based on local coordinates in each element, a general key addressing scheme can be created by encoding the refinement history of an element or node. There is a finite number of refinement rules which can be applied to an element. Each rule is assigned a number, and the key of an element of an adaptively refined mesh is composed by the number of its (grand-) parent elements of the initial coarse mesh and by the sequence of numbers of the refinement rules applied to the (grand-) parent elements.

3.3.6 Hashing

An efficient implementation of a key-based addressing scheme has to deal with the problem that the space of possible keys is huge and only a few of the keys are actually in use. In comparison to sorted lists and trees, where entries can be looked up by their respective keys, hash storage techniques are more efficient in the sense of a lower average computational complexity for the look-up operation. Hashing is a very general storage concept in computer science, see Knuth [186], often used in data base systems, but also in compilers or command interpreters. It is used to store and retrieve large amounts of data without relying on any special structure or layout of data. A related key-based addressing with hash storage is outlined by Williams [308].

A very general universe of all possible key values is mapped to an index space by a hash function f, see Figure 3.13. The index space consists of a finite number values, each of which represents a single memory cell. The cells can be organised in an ordinary vector in memory. An item with key k is stored in location $f(k)$ of the hash table. Several items with different keys $k_1 \neq k_2$ might be mapped to the same location $f(k_1) = f(k_2)$ which is called a collision and which has to be resolved. Collisions may happen because the hash function cannot be injective. Popular techniques to resolve these collisions are chaining (see Figure 3.13), linear probing and double hashing. The chaining strategy provides a linked list at each entry of the hash table, so that additional entries can be inserted. Under the assumption that the average length of the chains is small, the amount of work associated with linear searching in the linked lists is negligible. However, the memory requirements of this data structure depend on both the size of the hash table and the number of entries, while the performance primarily depends on the size of the hash table and the hash function f.

Other collision resolution techniques work on the hash table itself and avoid allocating additional memory. Linear probing means that if a location is filled already another location at a fixed distance is used. This process is iterated until an empty location is found. Table look-up is similar in first looking at $f(k)$ and probing linearly until the entry or an empty location is found. In the latter case, the entry is not present in the hash table. Efficiency of the hash table is given if the number of probing steps is small on the average. Hence the hash table has to be large enough so that a substantial number of locations is empty. Furthermore, it is more difficult to remove entries from the hash table based on probing than for a table based on chaining, where an entry can be removed from the linked list instead. This is because table look-up always requires a consistent hash table. An alternative to linear probing is double

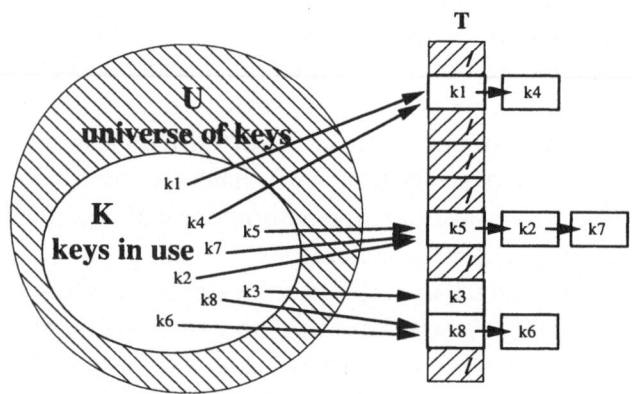

Figure 3.13. *Hash table, collision resolution with chaining.*

hashing, where the probing with a constant increment in case of a collision is replaced by a second hash function.

There are many different choices for a good hash function which is responsible for scattering the keys in use to the limited number of entries in the table. The performance of a hash table is usually estimated in a statistical setting, which depends on the quality of the hash function f. Random access is a constant time $\mathcal{O}(1)$ operation in the statistical mean, as long as there are enough empty slots in the table. Since a hash table of a suitable size is needed for efficiency, it may be necessary to re-size or enlarge a given table. Usually this requires creating a completely new table, since all entries have to be checked and probably placed elsewhere. In order to minimise the effort in re-sizing, which is an operation linear in the amount of data, the number of re-sizing steps has to be small. Without any further information, a strategy where the table is doubled in size gives a geometric growth and hence an overall constant amount of work per entry. For the efficiency of the hash table it is often useful to have hash table sizes of a large prime number, hence the size will be doubled only roughly. However, if the number of entries of the hash table can be estimated in advance, it will be even more efficient to create the hash table of an appropriate larger size instead of re-sizing.

In comparison to linked lists and trees, key-based addressing with hash storage is simple, efficient and requires very little memory. It provides a clean interface between the data container and algorithms on the data. Furthermore, keys can be advantageous for parallel execution, where pointers in a distributed memory environment are potentially dangerous and indexes only reference local arrays. Here keys provide a unique characterisation of nodes for the communication between processors and across memory boundaries.

3.3.7 Hierarchies of Meshes for Multilevel Methods

The difference in mesh storage between adaptively refined meshes and multilevel methods is that several meshes are needed at a time for multilevel methods. On structured meshes it is easy to extract coarser meshes from a given stored fine mesh. The nodes along one coordinate direction can be enumerated by $0, 1, \ldots, 2^l$ for dyadic refinement and coarsening. A nested sequence of coarser meshes on level i can be obtained by an iteration over the nodes by a larger increment of 2^{l-i}. The remaining nodes can be equivalently rearranged into a mesh with numbers $0, 1, \ldots, 2^i$. Hence no additional data structures are necessary to represent structured meshes for multilevel algorithms.

However, mesh coarsening for unstructured meshes is not immediately clear. General unstructured meshes cannot be transformed into coarser meshes by simple coarsening rules so that neighbour elements are agglomerated. Only algebraic multigrid methods are able to create coarser meshes based on several heuristics, either by a selection of coarse mesh nodes or by agglomeration. In general, these spaces cannot be represented as standard finite element discretisations on nested finite element spaces. Hence it is necessary for ordinary multilevel methods on adaptively refined meshes to store information on the coarsest initial mesh and on the refinement history, in order to use the nested finite element spaces.

Although it is not necessary to store the complete refinement history such as the sequence of all adaptively refined meshes, meshes on different levels and an interpolation between them should be extractable from the data structure. One way to organise this for unstructured meshes is tree data structures, where pointers refer to parent and children elements, see Figure 3.14. All elements of all meshes are stored in one tree as well as edges and nodes, see Rheinboldt and Mesztenyi [253] and Rüde [258]. A subset of the elements, nodes and edges forms one mesh. All nodes together with some of the elements form the finest mesh, where a solution is sought. Coarser level meshes of level l can be constructed as the union of the elements of tree level l. Alternatively, they can be chosen as the mesh which was created at refinement step l. The multigrid method and mesh refinement algorithms can be formulated to operate completely on trees with optimal complexity, see Rivara [255] and Leinen [197].

Hierarchical parent-children relations on edges can be stored. The storage of node hierarchies is less convenient, since there is a variable number of references for a single node from the coarsest mesh on which the node is created to

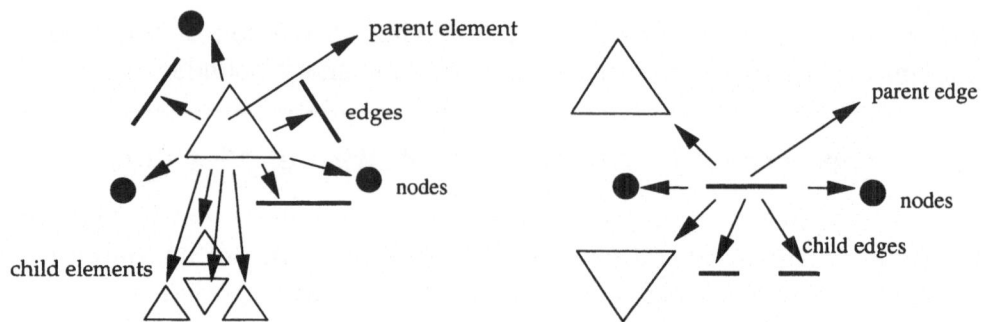

Figure 3.14. *An element and an edge data structure for the representation of unstructured meshes.*

the finest mesh in use. Additionally, nodes may exist as multiple instances on different levels or just as a single instance, which depends on the algorithm. Multiplicative algorithms require a node on each level, while an additive or a hierarchical basis algorithm uses the node at its coarsest level only. In a local multigrid method, a node has to be present on the coarsest mesh on which it is introduced. Furthermore, local neighbour nodes are needed on that level. Local multigrid may use a single node on several levels, but usually not on all possible levels as a standard multiplicative multigrid.

The tree data structures can be optimised for lower memory consumption by the elimination of some pointers. E.g., it will be sufficient to reference a single child node instead of all child nodes for red refinement, if neighbour links within each element reveal the remaining children elements. Furthermore, edge and face data structures may be eliminated, since there are many more edges and faces than nodes and elements in a large unstructured mesh, according to Euler's polyhedron formula. Sometimes pointers can also be eliminated if connected data structures occupy consecutive memory addresses, so that neighbours or children can be referenced by index arithmetic. However, it does not seem to be useful to throw away all interconnection information in favour of some search procedures, since the construction of coarse mesh discretisation and transfer operations should not require more than linear computational complexity.

In a key-based addressing scheme on an adaptively refined mesh, mesh hierarchies do not have to be stored explicitly. The construction of the keys allows one to reconstruct coarser mesh levels. The key-based addressing scheme is based on an initial coarse mesh, which is either a structured mesh or a small unstructured mesh. The sequence of refinement rules applied to an element are encoded in the key. Coarser mesh levels can be constructed in a natural way, starting with the initial coarse mesh and applying the encoded refinement rules only up to a certain level. This leads also to the appropriate inter-grid operators. Algorithmically, the coarser levels can also be constructed implicitly, where parent or children addresses are computed similar to Figure 3.12 and the elements or nodes are looked up in the hash table, see Figure 3.13.

Key-based addressing eliminates pointers. Hence an implementation requires less memory, is simpler, since update and maintenance of pointers is no longer needed, and the parallelisation is easier because key addresses are portable over memory boundaries.

3.3.8 Algorithmic Structures

The usual data structures for a PDE solver based on finite element or finite difference discretisations define the mesh, be it a structured, an unstructured or an adaptively refined mesh. The discretisation leads to a discrete differential operator, which can be stored in a sparse matrix data structure, or an algorithm, which computes the action of the operator on the fly. The algorithm would be based on the geometrical mesh and auxiliary data such as local stiffness matrices or coefficients of the operator. Furthermore, the solution, right hand side and auxiliary vectors may be stored in additional vector data structures. However, both matrices and vectors can also be distributed and stored within the geometrical mesh, see Figure 3.15 and [197, 257, 266, 317]. As mentioned before, the latter is less flexible and prohibits a clear distinction between discretisation and linear algebra.

Further data structures can be used to encapsulate algorithmic entities. Alternatively, procedural interfaces can be used for the same purpose, since only very little parameter data is associated directly to the algorithms. The basic building blocks of an adaptive parallel multilevel PDE solver are a method to discretise the PDE such as a matrix assembly or finite difference operator, methods to solve the equation systems such as linear and non-linear iterative solvers and a multilevel solver or preconditioner, methods for adaptive mesh refinement such as error estimators and mesh manipulation methods and parallel communication and load balancing methods, see also Figure 3.16.

In this chapter we have discussed several issues around the topic of adaptively refined meshes. The discretisation on adaptively refined meshes by finite element and finite different schemes, error estimation, mesh manipulation and suitable data structures were covered. These show a broad variety of possible choices for an adaptive code, including the type of mesh. Different mathemat-

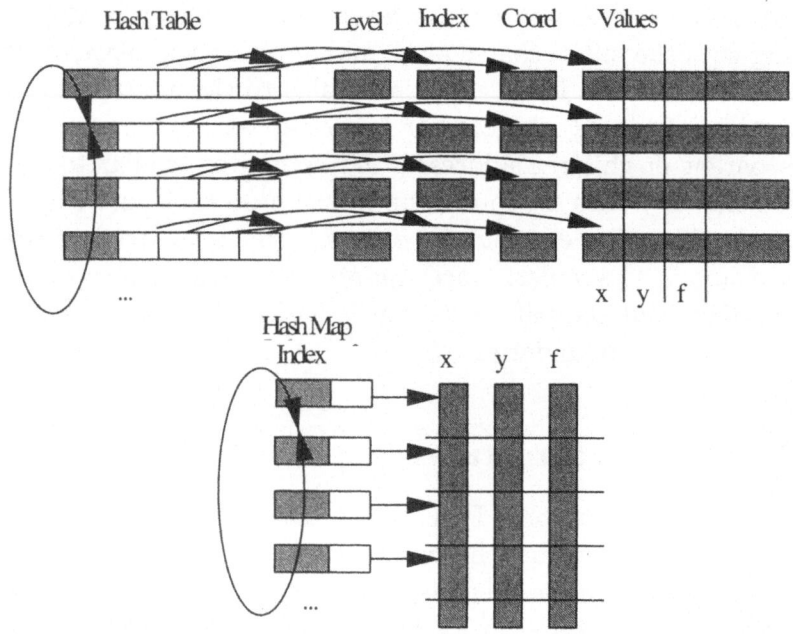

Figure 3.15. *Numerical data is stored within the mesh data structure (top) vs. numerical data is stored in external vector data structures and is indexed from the mesh (bottom).*

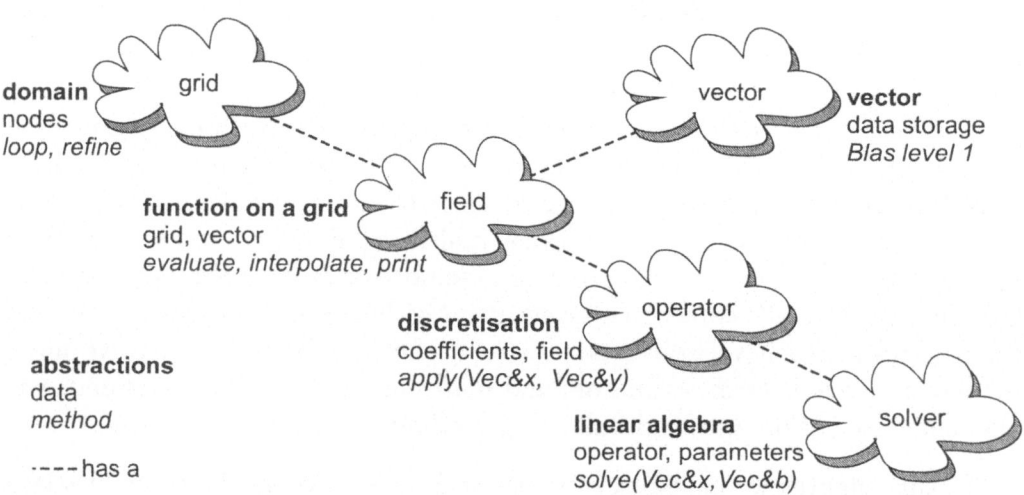

Figure 3.16. *Abstractions and data structures for a geometrical mesh, data and discretisations on the mesh.*

ical approaches lead to different error estimators. Under suitable assumptions, they give a reliable estimate on the size of the local error. Adaptive mesh refinement aims at an equilibration of a local error per work ratio, which is an equilibration of the local errors for local discretisation schemes. Together with assumptions on the strategy to select parts of the mesh for refinement, convergence of the adaptive refinement procedure can be proven, see Dörfler [105] and Cohen, Dahmen and DeVore [91].

A novel feature in the area of partial differential equations seems to be the key-based addressing. In comparison to other data structures developed for the efficient implementation of adaptive mesh refinement and multilevel algorithms on these mesh hierarchies, the abstraction of a key provides a simple interface between the mesh data storage and the algorithms on the mesh. This can be important in the case of parallelisation, when simplicity at the base level pays off at higher algorithmic levels. Note that such data structures are not needed if we either do not use a multilevel algorithm or if we do not handle adaptively refined meshes. Simpler data structures for both cases have been discussed in the previous section.

Chapter 4

Space-Filling Curves

When it comes to the parallelisation of a PDE solver, a data parallel approach is usually appropriate. We have to discuss the issue of data partitioning, which is a mesh partitioning in our case. So far we have discussed multigrid and multilevel algorithms in general and various aspects of adaptive mesh refinement, all on a sequential computer. In this chapter we will discuss the mesh partitioning or load balancing problem. We will do this by means of space-filling curves. While the general graph partitioning problem is NP-hard, meshes provide more information, namely the location of the nodes in space. Hence it seems to be more likely to construct mesh partitions of acceptable quality in polynomial time than for the pure graph partitioning problem. Nevertheless, we are looking for a heuristic which is linear in time, like the linear complexity multigrid solver, and which can additionally be parallelised as efficiently as the multigrid solver. This can indeed be done by space-filling curves, which enumerate all elements or nodes of a mesh. The partitioning algorithm simply chooses a complete interval of numbers which translates into a geometric domain.

The main focus of this chapter is on the analysis of this heuristic. Basically we are able to prove that the partitions are quasi-optimal. This means that the cut size of the graph is bounded by the optimal cut size times a constant, independently of the mesh size. This is true both for uniformly refined meshes and for certain types of adaptively refined meshes. Space-filling curves seem to be the first mesh partitioning heuristic for which such a property can be shown. Although some advanced graph partitioners may quickly produce slightly better cut sizes in numerical experiments, there seems to be no way to prove quasi-optimality for balanced partitions. Furthermore, even graph partitioners which are capable of the computation of un-balanced p-way partitions with good partition estimates are not incremental and have a higher sequential and parallel complexity than space-filing curve methods.

We are able to characterise adaptive meshes by the refinement algorithm and the solution type, which are partitioned by space-filling curves in a quasi-optimal way, and we introduce criteria which we call β- and γ-adaptive families of meshes. We consider a sequence of adaptively refined meshes, where the number of elements increases monotonically from one mesh to the next one. The β criterion describes the difference between regions with little refinement and regions with most refinement. An element is assumed to be subdivided into a number of elements on the next finer mesh. The factor β describes the quotient of an upper and a lower global bound for these numbers. For small differences, the estimates of uniformly refined meshes can almost be recovered. However, in higher dimensions d, or for steep adaptive mesh refinement, the estimates for β-refined meshes are not sharp enough. Hence we propose another criterion which describes upper and lower bounds for the refinement of an element not just for one refinement, but for the whole sequence of refinements. This way we are able to prove sharp estimates for different adaptive refinement procedures. Both popular adaptive mesh refinement heuristics and best n-term approximation meshes have this property, as long as the dimension of the sub-manifold which characterises the singular behaviour of the solution is bounded by some dimension dependent constant. This means that meshes for point singularities can always be partitioned quasi-optimally, and e.g. edge-singularities will usually be partitioned quasi-optimally only in dimensions higher than two. However, it is possible that even adaptive mesh refinement toward higher dimensional manifolds is partitioned in an optimal way, but counter examples show that such partitions may also have cut sizes proportional to the number of nodes.

Finally space-filling curve partitioning is also applied to sparse mesh discretisations, where it is shown that cut sizes are generally larger than for standard local discretisations. Furthermore, experimental results show that the estimated cut sizes are indeed observable in practice, which indicates that the constants involved are small.

We want to run the PDE solver on a parallel computer. We are interested in distributed memory computers and a large number of degrees of freedom. A data parallel approach to parallelisation is most appropriate where little data is transferred between the processors. Data is partitioned and each block of data is mapped to a processor. Partitioning data can be done in several ways. We need a method to do this at run-time, because the meshes are also created at run-time by the adaptive mesh refinement procedure. The data partitioning problem cannot be solved exactly, because in general it is NP-

hard, see Bui and Jones [76]. Heuristics either work top-down like recursive coordinate- or spectral-bisection (see Pothen, Simon and Liou [245]) or bottom-up like some graph agglomeration methods (see Karypis and Kumar [183]). However, we do not need a single partition of a single mesh, but a partition of a sequence of nested meshes for the multigrid method and a sequence of re-partitioning during adaptive mesh refinement. A generalisation of the single partition would be a fine grain *multilevel* partition, where each mesh level is cut into a large number of pieces. This results in a partition on each level suitable for a multilevel solver. Furthermore, we can agglomerate the pieces in different ways to accommodate changes due to adaptive mesh refinement without re-partitioning and changing the whole partition. We will explain this concept of a *multilevel* partition in the following chapter.

A natural representation of a multilevel partition of a mesh is an enumeration of the elements. Given an enumeration of the elements on all mesh levels, we can order them. On a certain mesh level, we can apply a recursive bisection based on the element order to form a partition onto p processors. Heuristically, the partitions should be connected with a small boundary to neighbour partitions. This is equivalent to some locality properties of the element enumeration. When some elements are added or deleted during adaptive mesh refinement, the partition changes accordingly. For a parallel multilevel algorithm the partitions of the different levels have to be related. This can be translated into the property of the enumeration, i.e., coarse elements and the related fine elements should have numbers which are close.

The enumeration defined by the multilevel partition describes a discrete space-filling curve of the domain Ω. The reverse statement also holds that each space-filling curve defines a multilevel partition. Hence we will review space-filling curves in general. Furthermore we are able to analyse the partitions defined by space-filling curves. This is remarkable, because there is little known analytically for graph partitioning heuristics. Usually, only experimental data is available for comparison of different methods. However, using some abstract properties of certain space-filling curves like Hölder continuity, it is possible to give estimates for large problem classes. In this way it is possible to show that under some assumptions, some space-filling curves give asymptotically optimal results, up to a constant.

4.1 Definition and Construction

First we shall define space-filling curves, sometimes also called self-avoiding walk.

Definition 4.1.

(a) *The term curve denotes the image of a continuous mapping f of the unit interval $I := [0, 1]$ to the compact domain $\overline{\Omega} \subset \mathbb{R}^d$.*

(b) *A curve is space-filling if it is surjective and Ω has a positive measure in $\mathbb{R}^d$.*

$$f : I \mapsto \Omega \subset \mathbb{R}^d, \qquad f \text{ continuous and surjective}. \qquad (4.1)$$

Mathematically, a curve is called space-filling if and only if the image of the mapping does have a classical positive d-dimensional measure, i.e. the curve fills up a whole domain. Hence the space-filling curve is surjective for a domain Ω of positive d-dimensional measure. In contrast to space-filling curves, injective curves are also called Jordan curves. Although Cantor proved the existence of bijective mappings between smooth, arbitrary, but finite dimensional manifolds in 1878, such mappings cannot be continuous. This was shown by Netto a year later.

Lemma 4.2. *A bijective mapping between two manifolds A, B, each with smooth boundary and both with different dimensions $m \neq n$ is discontinuous.*

The proof of Lemma 4.2 was given by Netto [221].

The discontinuous bijective mappings were initially also called *topological monsters*. However, if we allow for non-smooth boundaries, continuous space-filling Jordan curves exist: Osgood [227] constructed Jordan curves which fill a domain of positive measure with other space-filling curves as limiting arc. The boundary of the range of the mapping is not smooth but fractally shaped.

In the following, we restrict our attention to simple domains like the unit square or unit cube or other simple coverings of the domain of interest Ω. Hence Jordan curves cannot be space-filling for such domains, and we will drop the injectivity of the curves. The most prominent space-filling curves were described by Peano in 1890 [238] and Hilbert in 1891 [175]. Later on, more space-filling curves were constructed by Moore, Lebesgue, Sierpiński and many others, see also Sagan [261]. We will introduce these classical space-filling curves. Furthermore, we will construct special space-filling curves on general unstructured meshes. The space-filling curves can be used for data partitioning. Due to self similarity features of standard space-filling curves, multilevel partitions can also be constructed. We will derive estimates for such partitions which prove asymptotic optimality of the partitions. This is remarkable since such estimates do not exist for many other graph partitioning methods.

One of the oldest space-filling curves, the Hilbert curve, can be defined geometrically, see Hilbert [175]. If the interval I can be mapped to Q by a space-filling curve, then this must be true also for the mapping of four quarters (two dimensional case) of I to the four quadrants of Q, see Figure 4.1. Iterating this subdivision process while maintaining the neighbourhood relationships between the intervals leads to the Hilbert curve in the limit case. The mapping is defined by the recursive subdivision of the interval I and the square Q.

The construction begins with a generator template, which defines the order in which the four quadrants are visited. The template is applied in every quadrant and its sub-quadrants. By affine mappings and connections between the loose ends of the pieces of curves, the Hilbert curve is obtained. Note that the generator template is mirrored or rotated in some sub-quadrants.

There are basically two different Hilbert curves for the square modulo symmetries, see Figure 4.1 and Figure 4.2. An open and a closed space-filling curve can be constructed by the Hilbert curve generator maintaining both neighbourhood relations and the subdivision procedure. The closed curve is also called Moore curve, see Moore [215]. In three dimensions, the Hilbert curve is based on a subdivision into eight octants. It is no longer unique up to symmetry transformations. In fact there are 1536 different versions of Hilbert curves for the unit cube, see Alber and Niedermeier [3]. The first iterates of one of them are shown in Figures 4.3 and 4.4. A general scheme in order to computed Hilbert curves in arbitrary dimensions is given in Figure 4.5 and as an example the explicit cases in three dimensions in Figure 4.6. Note that computationally more efficient implementations are usually based on transformation tables like in Figure 4.7 instead of case statements. These tables can be generated automatically by the code described.

Some curves as intermediate results of iterative construction procedure, called discrete space-filling curves, are more interesting than the final Hilbert curve itself. The discrete curves are injective Jordan curves, but do not fill the domain Ω. They are also called self-avoiding walks. The final curve however is not self-avoiding. It would require further technical modifications to obtain an invertible mapping, which would not be continuous any more. We will prove the basic properties of the Hilbert curve we will need in the sequel.

Ueber die stetige Abbildung einer Linie auf ein Flächenstück.*)

Von

DAVID HILBERT in Königsberg i. Pr.

Peano hat kürzlich in den Mathematischen Annalen**) durch eine arithmetische Betrachtung gezeigt, wie die Punkte einer Linie stetig auf die Punkte eines Flächenstückes abgebildet werden können. Die für eine solche Abbildung erforderlichen Functionen lassen sich in übersichtlicherer Weise herstellen, wenn man sich der folgenden geometrischen Anschauung bedient. Die abzubildende Linie — etwa eine Gerade von der Länge 1 — theilen wir zunächst in 4 gleiche Theile 1, 2, 3, 4 und das Flächenstück, welches wir in der Gestalt eines Quadrates von der Seitenlänge 1 annehmen, theilen wir durch zwei zu einander senkrechte Gerade in 4 gleiche Quadrate 1, 2, 3, 4 (Fig. 1). Zweitens theilen wir jede der Theilstrecken 1, 2, 3, 4 wiederum in 4 gleiche Theile, so dass wir auf der Geraden die 16 Theilstrecken 1, 2, 3, ..., 16 erhalten; gleichzeitig werde jedes der 4 Quadrate 1, 2, 3, 4 in 4 gleiche Quadrate getheilt und den so entstehenden 16 Quadraten

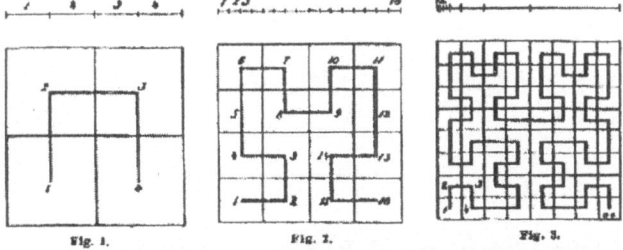

Fig. 1.   Fig. 2.   Fig. 3.

werden dann die Zahlen 1, 2 ... 16 eingeschrieben, wobei jedoch die Reihenfolge der Quadrate so zu wählen ist, dass jedes folgende Quadrat sich mit einer Seite an das vorhergehende anlehnt (Fig. 2). Denken wir uns dieses Verfahren fortgesetzt — Fig. 3 veranschaulicht den

*) Vergl. eine Mittheilung über denselben Gegenstand in den Verhandlungen der Gesellschaft deutscher Naturforscher und Aerzte. Bremen 1890.
**) Bd. 36, S. 157.

30*

Figure 4.1. *Historic drawing and construction of a Hilbert space-filling curve [175]. Template and two refinement steps.*

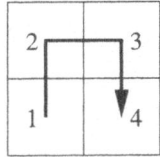
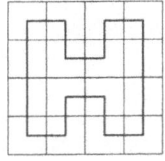
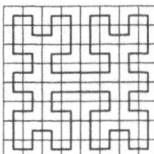

Figure 4.2. *The closed Hilbert-Moore space-filling curve starting with the same template.*

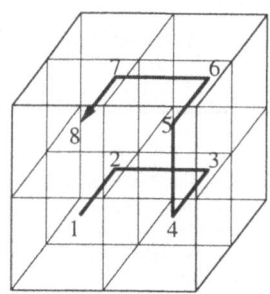 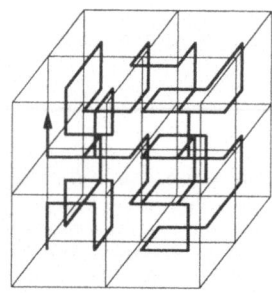

Figure 4.3. *A three-dimensional Hilbert curve. Template and first refinement step.*

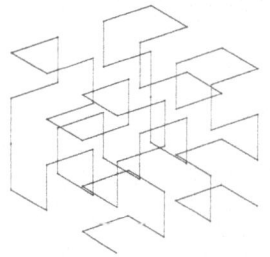 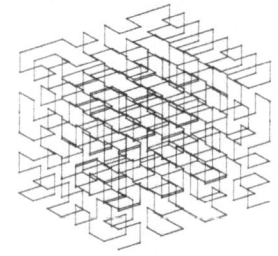 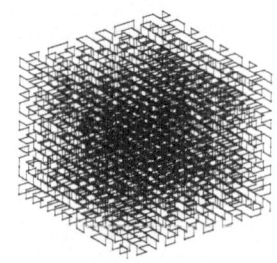

Figure 4.4. *A three-dimensional Hilbert curve. A sequence of refinement steps.*

Lemma 4.3. *The Hilbert mapping $I \mapsto [0,1]^d = Q$ is a non-injective, nowhere differentiable for $d > 1$, Hölder-continuous space-filling curve. The mapping is*

(a) *continuous, i.e. a curve,*

(b) *surjective, i.e. space-filling,*

(c) *not injective, i.e. not self-avoiding,*

(d) *nowhere differentiable and*

(e) *Hölder continuous with exponent $1/d$:*

$$\|f(x) - f(y)\|_2 \leq C \sqrt[d]{|x - y|} \text{ for all } x, y \in I \qquad (4.2)$$

```
keytype coord2key(int x[d], Box box) {
   keytype s, c = 0;
   int l[d] = {0,0,...};
   int h[d] = {1,0,...};
   for (int i=0; i<level; i++) {
      calculate coordinates x relative to box;
      calculate affine mapping T from l, h to {0,0,...}, {1,0,...};
      transform x by T;
      calculate s, new l, new h;
      calculate child of box;
      transform l, h back by T⁻¹;
      c = c * 2ᵈ + s;
   }
   return c;
}
```

Figure 4.5. *Pseudo-code for the construction of a d-dimensional Hilbert curve. Transform a coordinate based representation* x *to the curve index* c.

```
if (x[0]==0) {                                if (x[1]==0) {
  if (x[1]==0) {                                if (x[2]==0) { // 1 0 0
    if (x[2]==0) { // 0 0 0                        s = 7; l = {1, 0, 1}; h = {1, 0, 0};
      s = 0; l = {0, 0 ,0}; h = {0, 0, 1};      } else {        // 1 0 1
    } else {        // 0 0 1                        s = 6; l = {1, 1, 0}; h = {1, 0, 0};
      s = 1; l = {0, 0, 0}; h = {0, 1, 0};      }
    }                                           } else {
  } else {                                        if (x[2]==0) { // 1 1 0
    if (x[2]==0) { // 0 1 0                         s = 4; l = {0, 1, 1}; h = {1, 1, 1};
      s = 3; l = {0, 1, 1}; h = {1, 1, 1};      } else {        // 1 1 1
    } else {        // 0 1 1                        s = 5; l = {1, 1, 0}; h = {1, 0, 0};
      s = 2; l = {0, 0, 0}; h = {0, 1, 0};      }
    }                                           }
  }                                           }
} else {
```

Figure 4.6. *Part of the pseudo-code for the construction of a three-dimensional Hilbert curve. The eight cases represent the eight cells of the template of the curve. The code can be plugged into Figure 4.5 where* s, l *and* h *are to be calculated.*

Proof of Lemma 4.3.

(a) We will directly show the stronger statement (e) which implies (a).

(b) Given an arbitrary point $x \in Q$, a nested sequence of sub-quadrants Q_i of Q with $x \in Q_i$ can be found. The limit of the sequence is a single point $\lim_{i \to \infty} Q_i = \{x\}$. A point $t_i \in I$ is mapped to each sub-quadrant

```
const int State[24][8] =
  { { 16,   8,   8,   6,   6,  11,  11,  19 },   // 0
    { 18,  10,  10,   7,   7,   9,   9,  17 },   // 1
    { 12,  17,  17,   4,   4,  18,  18,  15 },   // 2
    { 14,  19,  19,   5,   5,  16,  16,  13 },   // 3
    {  9,  20,  20,   2,   2,  23,  23,  10 },   // 4
    { 11,  22,  22,   3,   3,  21,  21,   8 },   // 5
    { 21,  13,  13,   0,   0,  14,  14,  22 },   // 6
    { 23,  15,  15,   1,   1,  12,  12,  20 },   // 7
    {  0,  16,  16,  14,  14,  21,  12,   5 },   // 8
    {  4,  20,  20,  15,  15,  17,  17,   1 },   // 9
    { 18,   1,   1,  12,  12,   4,   4,  23 },   // 10
    { 22,   5,   5,  13,  13,   0,   0,  19 },   // 11
    { 17,   2,   2,  10,  10,   7,   7,  20 },   // 12
    { 21,   6,   6,  11,  11,   3,   3,  16 },   // 13
    {  3,  19,  19,   8,   8,  22,  22,   6 },   // 14
    {  7,  23,  23,   9,   9,  18,  18,   2 },   // 15
    {  0,   8,   8,  22,  22,  13,  13,   3 },   // 16
    {  2,  12,  12,  23,  23,   9,   9,   1 },   // 17
    { 10,   1,   1,  20,  20,   2,   2,  15 },   // 18
    { 14,   3,   3,  21,  21,   0,   0,  11 },   // 19
    {  9,   4,   4,  18,  18,   7,   7,  12 },   // 20
    { 13,   6,   6,  19,  19,   5,   5,   8 },   // 21
    {  5,  11,  11,  16,  16,  14,  14,   6 },   // 22
    {  7,  15,  15,  17,  17,  10,  10,   4 } };  // 23

const int Curve[24][8] =
  { {  0,   7,   1,   6,   3,   4,   2,   5 },   // 0
    {  7,   0,   6,   1,   4,   3,   5,   2 },   // 1
    {  3,   4,   0,   7,   2,   5,   1,   6 },   // 2
    {  4,   3,   7,   0,   5,   2,   6,   1 },   // 3
    {  1,   6,   2,   5,   0,   7,   3,   4 },   // 4
    {  6,   1,   5,   2,   7,   0,   4,   3 },   // 5
    {  2,   5,   3,   4,   1,   6,   0,   7 },   // 6
    {  5,   2,   4,   3,   6,   1,   7,   0 },   // 7
    {  0,   1,   3,   2,   7,   6,   4,   5 },   // 8
    {  7,   6,   4,   5,   0,   1,   3,   2 },   // 9
    {  3,   0,   2,   1,   4,   7,   5,   6 },   // 10
    {  4,   7,   5,   6,   3,   0,   2,   1 },   // 11
    {  1,   2,   0,   3,   6,   5,   7,   4 },   // 12
    {  6,   5,   7,   4,   1,   2,   0,   3 },   // 13
    {  2,   3,   1,   0,   5,   4,   6,   7 },   // 14
    {  5,   4,   6,   7,   2,   3,   1,   0 },   // 15
    {  0,   1,   7,   6,   3,   2,   4,   5 },   // 16
    {  7,   6,   0,   1,   4,   5,   3,   2 },   // 17
    {  3,   0,   4,   7,   2,   1,   5,   6 },   // 18
    {  4,   7,   3,   0,   5,   6,   2,   1 },   // 19
    {  1,   2,   6,   5,   0,   3,   7,   4 },   // 20
    {  6,   5,   1,   2,   7,   4,   0,   3 },   // 21
    {  2,   3,   5,   4,   1,   0,   6,   7 },   // 22
    {  5,   4,   2,   3,   6,   7,   1,   0 } };  // 23

keytype coord2key(int x[3]) {
    int lh=0;
    keytype c = 0;
    for (int i=level-1; i>=0; i--) {
      int box = ((x[0] >> i) & 1) |
                ((x[1] >> (i-1)) & 2) | ((x[2] >> (i-2)) & 4);
      c  = Curve[lh][box] + c * 8;
      lh = State[lh][box];
    }
    return c;
}
```

Figure 4.7. *Complete pseudo-code for the construction of a three-dimensional Hilbert curve based on tables. Only half of the 24 affine transformations and states* lh *are actually used.*

Q_i and the limit point $t = \lim_{i \to \infty} t_i$ is mapped to x. The mapping is surjective.

(c) We observe that points $x \in Q$ at a common boundary of a quadrant may have several distinct inverse-images. This can be demonstrated easily by looking at the centre point $(1/2, 1/2, \ldots) \in Q$, which is contained in the image of all 2^d quadrants. Hence several distinct points on I are mapped to this point $x \in Q$ and the mapping f is not injective.

(d) Given an arbitrary point $t \in I$ on the interval. On a mesh of $2^n \times 2^n \times \ldots$ quadrants of side length 2^{-n}, we look for a point t_n in the neighbourhood with $|t - t_n| \le (3^d + 1)2^{-dn}$ such that the quadrants to which t and t_n

are mapped do not share a common boundary. The quadrants are at least 2^{-n} apart. Such a point t_n does exist, because only $(3^d - 1)$ other quadrants may share a common boundary with any given quadrant and we allow for a distance of $(3^d + 1)$ quadrants. Now we have

$$\frac{\|f(t) - f(t_n)\|_2}{|t - t_n|} \geq \frac{2^{(d-1)n}}{1 + 3^d},$$

which is not bounded for $n \to \infty$. Hence the mapping f is not differentiable in t. Furthermore, a Hölder exponent larger than $1/d$ is not possible.

(e) Given two arbitrary points $t_1, t_2 \in I$ on the interval at a distance $k = |t_2 - t_1|$. We choose the mesh of sub-quadrants of side length 2^{-n} such that the volume of each sub-quadrant is at least k, that is $2^{-d(n+1)} < k \leq 2^{-dn}$. The mapping f cycles through a complete sub-quadrant before entering a new one. Hence, the image of $f([t_1, t_2])$ is contained in at most two adjacent sub-quadrants. The maximum Euclidean distance within the two sub-quadrants is $2^{-n}\sqrt{d+3}$ calculated by a distance 2^2 along one axis and 1 along the other $(d-1)$ axes times the size 2^{-n}. Hence $\|f(t_2) - f(t_1)\|_2 \leq 2^{-n}\sqrt{d+3} \leq 2\sqrt{d+3}\sqrt[d]{|t_2 - t_1|}$. $\square$

Further proofs of (a), (b) and (d) can also be found in some textbooks including Sagan [261, chapter 2]. Results for special cases of discrete curves in two or three dimensions (e), some of them with tighter bounds C may be found, sometimes in a different context, in Stout [288], Chochia, Cole and Heywood [87], Gotsman and Lindenbaum [130], Alber and Niedermeier [3], and Hungershöfer and Wierum [180].

Algorithmically, the two-dimensional Hilbert curve mapping of a point $x \in [0, 1]$ can be expressed as follows. We assume that the number x is given in quaternary representation as $0_4.q_1q_2q_3\ldots$.

$$f(0_4.q_1q_2q_3q_4\ldots) = \mathcal{H}_{q_1} \circ \mathcal{H}_{q_2} \circ \mathcal{H}_{q_3} \circ \mathcal{H}_{q_4} \ldots \begin{pmatrix} 0 \\ 0 \end{pmatrix}$$

with affine mappings $\mathcal{H}_0$, $\mathcal{H}_1$, $\mathcal{H}_2$, $\mathcal{H}_3$ defined as

$$\mathcal{H}_0\begin{pmatrix} x \\ y \end{pmatrix} = \begin{pmatrix} 0 & \frac{1}{2} \\ \frac{1}{2} & 0 \end{pmatrix}\begin{pmatrix} x \\ y \end{pmatrix},$$

$$\mathcal{H}_1\begin{pmatrix} x \\ y \end{pmatrix} = \begin{pmatrix} \frac{1}{2} & 0 \\ 0 & \frac{1}{2} \end{pmatrix}\begin{pmatrix} x \\ y \end{pmatrix} + \begin{pmatrix} 0 \\ \frac{1}{2} \end{pmatrix},$$

$$\mathcal{H}_2\begin{pmatrix} x \\ y \end{pmatrix} = \begin{pmatrix} \frac{1}{2} & 0 \\ 0 & \frac{1}{2} \end{pmatrix}\begin{pmatrix} x \\ y \end{pmatrix} + \begin{pmatrix} \frac{1}{2} \\ \frac{1}{2} \end{pmatrix},$$

$$\mathcal{H}_3\begin{pmatrix} x \\ y \end{pmatrix} = \begin{pmatrix} 0 & -\frac{1}{2} \\ -\frac{1}{2} & 0 \end{pmatrix}\begin{pmatrix} x \\ y \end{pmatrix} + \begin{pmatrix} 1 \\ \frac{1}{2} \end{pmatrix}.$$

The related mapping of the discrete Hilbert curve can be obtained by truncation. Given the number $x = 0_4.q_1q_2\ldots q_n$, the corresponding position on the Hilbert curve of fineness 4^n can be computed by

$$f_n(0_4.q_1q_2\ldots q_n) = \mathcal{H}_{q_1} \circ \mathcal{H}_{q_2} \circ \ldots \circ \mathcal{H}_{q_n}\begin{pmatrix} 0 \\ 0 \end{pmatrix}.$$

Similar constructions are possible for higher dimensions d. Alternatively, the two dimensional Hilbert curve can also be constructed with complex numbers representing the affine mappings $\mathcal{H}_q$ in the real and imaginary part.

The Peano space-filling curve can be constructed by a subdivision scheme of the square Q into nine sub-squares of side length $1/3$, see Figure 4.8. The generator can be given by the order of the nine squares. In contrast to the Hilbert curve, there are 272 different curves up to similarities, which maintain neighbourhood relations. Basically, two different templates exist. We call the first one in Figure 4.8 Peano template and the second one (Meander-like) Peano-Wunderlich template in Figure 4.9. In addition to different templates, two orientations are possible for a curve which enters a square at one corner and leaves it at the opposite corner. This leads to the large number of different Peano curves. Of course, many more different higher dimensional Peano curves can be defined as well, by subdivision into 3^d cubes.

The construction of the Sierpiński curve, see Sierpiński [274], can be explained more easily by a triangulation than by a template like the Hilbert or Peano curves. We start with four triangles as in Figure 4.10 (left). The triangles are numbered, which we indicate by a curve that connects the centres of gravity of the triangles. The triangulation is refined by successive bisection of the longest edge. The numbering is changed in such a way that the child triangles substitute the single parent triangle, maintaining neighbourhood relations. The final Sierpiński curve is in fact equivalent to the limit of an alternative

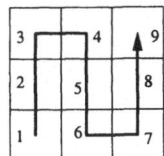 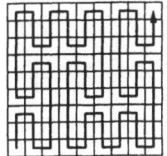 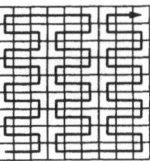

Figure 4.8. *The Peano curve: construction template (left) and two different Peano curves.*

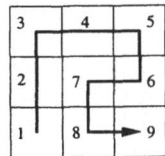 

Figure 4.9. *Meander-type Peano-Wunderlich curves: construction template (left) and a Peano curve.*

construction called H-index proposed by Niedermeier, Reinhardt and Sanders [223], see Figure 4.11. This can be seen roughly if one compares Figures 4.10 and 4.11 carefully. The equivalence was observed recently by Ellerbrake [114].

Lemma 4.4. *The Peano and Sierpiński mappings are non-injective, nowhere differentiable for $d > 1$, Hölder-continuous with exponent $1/d$ space-filling curves.*

The Proof of Lemma 4.4 is analogous to the proof of Lemma 4.3 with slightly different constants.

Tighter bounds C of the Hölder continuity for the two dimensional case of the Peano curve were derived by Garsia [124] and for the H-index which is equivalent to the Sierpiński curve by Chochia and Cole [86] and by Niedermeier, Reinhardt and Sanders [223]. Note that the Sierpiński curve appears to have the smallest constant C of all known self-similar space-filling curves.

The Sierpiński curve was defined originally on a single triangle. However, the curve can be generalised to arbitrary unstructured triangulations. We start with an arbitrary initial curve on the coarsest mesh defined by a numbering of the elements. A finite set of refinement rules, see e.g. [26, 41], is used to subsequently construct finer meshes and the curve is refined accordingly. Whenever an element is substituted by a set of smaller elements covering the same area,

 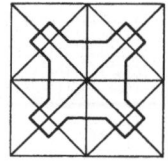 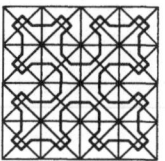

Figure 4.10. *Construction of the Sierpiński space-filling curve by a triangulation of the square. Three refinement steps are shown.*

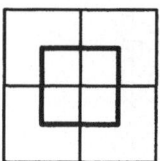 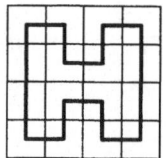 

Figure 4.11. *Construction of the H-Index which is equivalent to the Sierpiński curve in the limit. Template and two refinement steps.*

the curve is substituted by a new curve that cycles through these elements exactly where the original element was removed. This process definenes *generalised* Sierpiński curves on a sequence of meshes of polyhedral elements which is created by element refinement and which element sizes all tend to zero. Note that the curve is a space-filling curve, but need not be a self-avoiding walk any longer.

Note that a generalised Sierpiński curve is space-filling in the limit case of zero element sizes. Triangles can be bisected or cut into four similar smaller triangles, tetrahedra into two or eight smaller tetrahedra, see Figure 4.12. There might be impurities in the final generalised Sierpiński curve if the discrete curve links triangles over a common point instead of a common edge. This can be caused by the initial triangulation, which is a property of the graph of the mesh, or by the mesh refinement, if it is one of the bisection rules or adaptive mesh refinement. Similar space-filling curves can be constructed in the three-dimensional case for tetrahedral meshes. Here, impurities such that the curve does not cross triangle-faces every time from one tetrahedron to the next do occur necessarily with any regular or sufficiently general adaptive mesh refinement procedure. Some examples are shown in Figure 4.13 for regular mesh refinement with triangles and tetrahedra and in Figure 4.14 with adaptive mesh refinement.

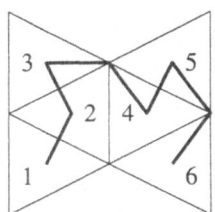 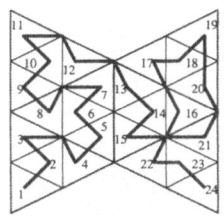 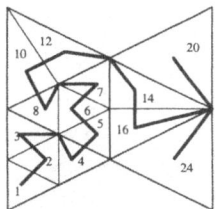

Figure 4.12. *Construction of a generalised Sierpiński-type space-filling curve. Initial triangulation (left), one uniform refinement step (middle) and one adaptive refinement step (right, red-green refinement). The curve is parametrised as $I = [0, 1]$ and the numbers of the triangles have to be divided by 6 (left) respectively 24 (middle and right).*

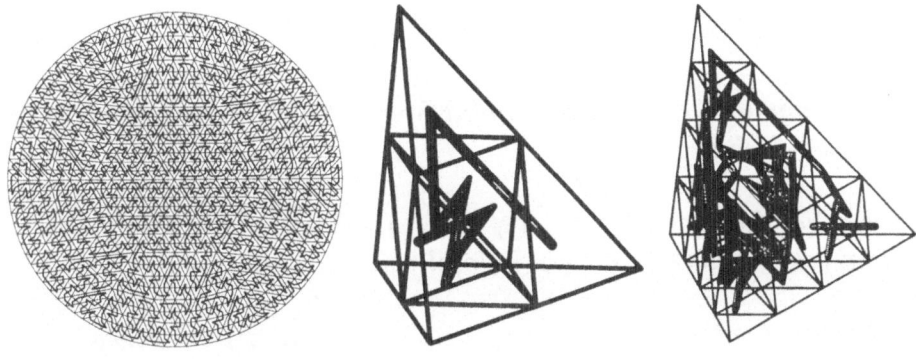

Figure 4.13. *Generalised Sierpiński-type space-filling curves defined on meshes with uniform mesh refinement.*

Definition 4.5. *We define a quasi-uniform refinable family T of meshes as a nested sequence of meshes τ_i which covers a domain Ω. Each mesh consists of a finite number of non-overlapping polygonally shaped open elements $e_{i,j} \in \tau_i$ by the conditions:*

(a) *$\overline{\bigcup_j e_{i,j}} = \Omega$ and $e_{i,j} \cap e_{i,k} = \emptyset$ for $j \neq k$, (grid condition),*

(b) *the element size has uniformly limit zero, $\lim_{i \to \infty} \max_j \operatorname{vol}(e_{i,j}) = 0$ (T is dense),*

(c) *either $e_{i,j} \in \tau_{i+1}$ or a set K of smaller elements exists with $\overline{e_{i,j}} = \overline{\bigcup_{k \in K} e_{i+1,k}}$ (nestedness),*

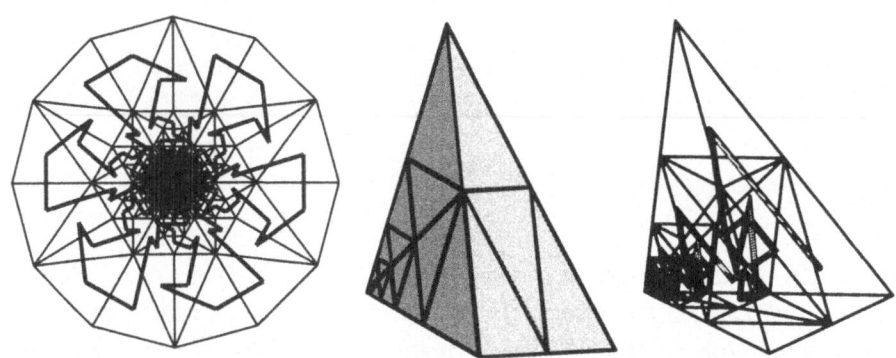

Figure 4.14. *Generalised Sierpiński-type space-filling curves on triangular and tetrahedral meshes with steep adaptive mesh refinement and arithmetic progression.*

(d) *the cardinality $|K| \leq c_c$ is bounded from below independent of i, j (finite number of children),*

(e) *the ratio of volumes $\mathrm{vol}(e_{i+1,k})/\mathrm{vol}(e_{i,j}) \geq c_{\min} > 0$ for $k \in K$ is bounded from below independent of i, j (local volume condition),*

(f) *all interior angles of the element $e_{i,j}$ are bounded from below $\geq \alpha_0 > 0$ independent of i, j (angle condition),*

(g) *the number of vertices of the element $e_{i,j}$ is bounded by $\leq c_v$ (vertex condition).*

Many meshes used in the FEM will be quasi-uniform refinable, if we consider the limit case of step size $h \to 0$: of course uniformly refined meshes, starting on the unit square or an arbitrary unstructured mesh, fall into this category. Furthermore, meshes created by adaptive refinement by quad-tree or oct-tree refinement, with hanging nodes or some closure rules and any stable bisection schemes for triangle and tetrahedron meshes, including red-green refinements, if we temporarily remove the green closures, are all quasi-uniform refinable. Even meshes consisting of different element types like mixtures of tetrahedra, hexahedra, pyramids and prisms can be used for quasi-uniform refinable meshes. We do have a rather general class of geometric meshes. The class of meshes can be generalised to iso-parametric elements, where each polygonal shaped element is transformed by some diffeomorphisms.

Definition 4.6. *A quasi-uniform mesh T of a domain Ω is defined as a finite set of non-overlapping polyhedrally shaped elements e_j, which cover the domain and have the following properties:*

(a) $\overline{\bigcup_j e_j} = \Omega$ and $e_j \cap e_k = \emptyset$ for $j \neq k$, (grid condition),

(b) the variation of element volumes is bounded from above and from below, (quasi-uniform),

(b) all interior angles of the elements $e_{i,j}$ are bounded from below $\geq \alpha_0 > 0$ independent of i, j and the ratio of circumcircle diameter and inscribed circle diameter is bounded independent of i, j (minimum angle and shape condition),

(c) the number of vertices of the element $e_{i,j}$ is bounded by $\leq c_v$ (vertex condition).

For the analysis of generalised Sierpiński curves on quasi-uniform refinable families of meshes we need the following construction of a quasi-uniform mesh with bounded variation of element sizes. The term quasi-uniform only makes sense for a family of meshes which share the same constants. A single mesh with a finite number of elements does fulfil the requirements of course, since the variation of element sizes is finite.

Lemma 4.7. *Given a quasi-uniform refinable family of meshes T, for any volume v small enough we can construct a new quasi-uniform mesh τ with element volumes $vc_{min} \leq \text{vol}(e) \leq v$ for all $e \in \tau$ and the same constants c_v and c_α. A new quasi-uniform refinable family $\hat{T}$ can be constructed by a sequence of quasi-uniform meshes τ.*

The Proof of Lemma 4.7 is constructive. We start with the coarse mesh $\tau := \tau_0$. An element $e \in \tau$ is refined as long as its volume is larger than v. The volume of an element is reduced at each refinement step by a factor between c_{min} and $1 - c_{min}$ with $1 - c_{min} < 1$. Hence a finite number of refinement steps is sufficient and only a finite number of elements is created. The volume of an element in τ is between $c_{min}v$ and v by construction and by property (e). We can construct a nested sequence of quasi-uniform meshes by choosing the volume v_n for the construction of mesh $\hat{\tau}_n$ as $v_n = (1 - c_{min})^n$. $\square$

Now we can show that the number of neighbour elements is bounded for quasi-uniform meshes which we will need for the continuity of generalised

Sierpiński curves. However, this property does not hold for meshes in a quasi-uniform refinable family in general: just think of a non-conforming mesh where two macro elements are refined at a different rate and a large element meets with an increasing number of small elements at an edge.

Lemma 4.8. *The number of neighbour elements of an element $e_{i,j}$ on a mesh τ_i of a family of quasi-uniform meshes T is bounded independent of i, j.*

Proof of Lemma 4.7. Due to the minimum angle condition (f) only a limited number c_1 of elements meet at a single point. The bounded variation of element volumes together with the shape condition and c_v gives upper and lower bounds for edge lengths and face sizes. Hence only a bounded number c_2 of elements may meet at an element's edge or face. Multiplying both constants c_1 and c_2 by the number of vertices, edges and faces bounded by c_v gives an upper bound for the number of neighbour elements. $\square$

Definition 4.9.

(a) *The term* parametrisation by volume *of space-filling curves $f : I \mapsto \Omega$ shall be defined as a parametrisation, such that the volume of the image $F([t_0, t_1])$ of a sub-interval $[t_0, t_1] \subset I$ is proportional to the length of the sub-interval*

$$\lambda(f([t_0, t_1])) \; = \; \lambda(f(I))|t_1 - t_0| \,. \tag{4.3}$$

(b) *The term* parametrisation by element *of a space-filling curve is defined by the limit process of the sequence of meshes τ on which the construction of f is based. The discrete curve on a mesh $T \in \tau$ of n elements is parametrised such that the interval $(i/n, (i+1)/n)$ is mapped to element number i.*

The space-filling curves defined so far, namely the Peano, Hilbert and Sierpiński curves, are parametrised by volume and at the same time also parametrised by element. Furthermore, it is possible to construct the generalised Sierpiński curves in such a way that they are parametrised by volume: the parametrisation of a coarser mesh is maintained on finer meshes and the interval on an element is divided into sub-intervals proportional to the volume of the sub-elements. We will see in the section on adaptive mesh refinement that such curves need not be parametrisable by element.

Lemma 4.10. *A generalised Sierpiński mapping on a quasi-uniform refinable family of meshes which is parametrised by volume is nowhere differentiable for $d > 1$. It is a space-filling curve. The mapping is*

(a) *surjective,*

(b) *not injective,*

(c) *nowhere differentiable and*

(d) *Hölder continuous with exponent $1/d$ (Equation (4.2)).*

Proof: Lemma 4.10 is a generalisation of Lemma 4.4. The constant C of Equation (4.2) depends on $\alpha_0, c_{\min}, C_m$ and the size of the minimum and maximum element of the initial mesh τ_0. The Sierpiński curve f maps the unit interval I to the domain Ω spanned by τ_0 and all subsequent meshes.

(a) Given an arbitrary point $x \in \Omega$, the point can be found in at least one element T. Furthermore x can be found in a whole sequence of nested sub-elements $T_i \subset T$ with $x \in T_i$. The limit of the sequence is a single point $\lim_{i \to \infty} T_i = \{x\}$. A point $t_i \in I$ is mapped to each element T_i and the limit point $t = \lim_{i \to \infty} t_i$ is mapped to x. The mapping is surjective.

(b) Every point x on the common edge of two elements T_1, T_2 can be found in two disjoint nested sequences of sub-elements with inverse-images $t_1, t_2 \in I$ such that $f(t_1) = f(t_2) = x$. The points t_1 and t_2 coincide only if the curve crosses the edge between T_1 and T_2 exactly at x. In this case we choose a different x on the edge. Finally we have $t_1 \neq t_2$ and $f(t_1) = f(t_2)$.

(c) Given an arbitrary point $t \in I$ on the interval. We construct a quasi-uniform mesh τ_n with element volumes between 2^{-dn} and $c_{\min} 2^{-dn}$ by Lemma 4.7. The image of t lies in the element T_n of the mesh τ_n. The element has a maximum number of neighbours c_α by Lemma 4.8. Hence we can find a point $t_n \in I$ with distance $|t - t_n| < (c_\alpha + 2)2^{-dn}$ and an image in element $\hat{T}_n$ of τ_n such that both elements do not share a common boundary. Due to the shape condition, the inscribed circle diameter of any element in τ_n is at least $C_\alpha \sqrt[d]{\text{volume}}$. Hence, the images of t and t_n are at least a distance $C_\alpha \sqrt[d]{c_{\min}} 2^{-n}$ apart. We have

$$\frac{\|f(t) - f(t_n)\|_2}{|t - t_n|} \geq 2^{(d-1)n} \frac{C_\alpha \sqrt[d]{c_{\min}}}{c_\alpha + 2},$$

which is unbounded for $n \to \infty$. The Hölder exponent cannot be larger than $1/d$.

(d) Given two arbitrary points $t_1, t_2 \in I$ on the interval at a distance $k = |t_2 - t_1|$. We construct a quasi-uniform mesh τ_n with element volumes between $k/c_{\min}$ and k by Lemma 4.7. The curve cycles through one element completely before it enters a new one. The interval $[t_1, t_2]$ is mapped to at most two adjacent elements, otherwise the element would be smaller than k. The maximum Euclidean distance within both elements is bounded by twice the circumcircle diameter, which is bounded by $C_\alpha \sqrt[d]{\text{volume}}$ due to the shape condition and c_v. The distance is at most $\|f(t_2) - f(t_1)\|_2 \leq 2C_\alpha \sqrt[d]{|t_2 - t_1|/c_{\min}}$. $\square$

There are two basic differences of the Lebesgue space-filling curve compared to the previous curves: this curve is differentiable almost everywhere, while the previous mentioned curves are not differentiable anywhere. The Hilbert, Peano and Sierpiński curves are self-similar, a feature they share with fractals: sub-intervals of the unit interval I are mapped to curves of similar structure as the original curve. However, the Lebesgue curve is not self-similar. It can be defined on the Cantor set Γ. This set is defined by the remainder of the interval I, after successively one third $(1/3, 2/3)$ has been removed. Any element $x \in \Gamma$ of the Cantor set can be represented as a number in ternary expansion, $0_3.t_1t_2t_3\ldots$ where all digits t_i are zeros and twos, i.e. $t_i \in \{0, 2\}$, because the ones have been removed by the construction of the Cantor set. The mapping of the Lebesgue curve, see Lebesgue [195], is defined by

$$f(0_3.(2d_1)(2d_2)(2d_3)\ldots) := \begin{pmatrix} 0_2.d_1d_3d_5\ldots \\ 0_2.d_2d_4d_6\ldots \end{pmatrix} \text{ with binary digits } d_i \in \{0, 1\} .$$

The function f defined on the Cantor set Γ is extended to the unit interval I by linear interpolation. The generator template of the Lebesgue curve looks like that in Figure 4.15. Although constructed by digit shuffling or *bit interleaving*, the curve is continuous and a space-filling curve. The order of the quadrants imposed by the generator of the curve can be found in the depth-first traversal of oct-trees and related algorithms.

Lemma 4.11. *The Lebesgue mapping is a non-injective almost everywhere differentiable Hölder-continuous with exponent $\frac{1}{d\log_2 3}$ space-filling curve.*

Proof of Lemma 4.11.
The mapping f is surjective, because $f(\Gamma) = Q$ already and it is not injective, because the mapping restricted to Γ is not injective. This can be shown

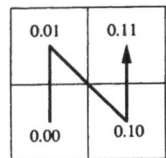 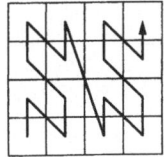 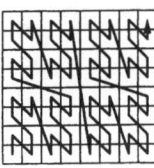

Figure 4.15. *The construction of the Lebesgue curve. Three successive refinement steps.*

analogously to Lemma 4.4(a) and (b). The mapping f restricted to $I \setminus \Gamma$ is affine, hence differentiable. It is differentiable on I with the exception of the set Γ of measure zero, i.e. almost everywhere. It remains to be shown that the mapping f is Hölder continuous.

Given a point $x \in I$. If $x \in I \setminus \Gamma$, an open neighbourhood $U(x)$ in I will belong to $I \setminus \Gamma$. The mapping f is affine in $U(x)$, hence Hölder continuous. Let us assume that $x \in \Gamma$. Any affine continuation of f left or right of x is Hölder continuous in a neighbourhood. Given another point $t \in \Gamma$ with $|t - x| < \frac{1}{3^{dn}}$, the ternary representation of both points coincides for the first dn digits. Hence the binary representation of their images $f(x)$ and $f(t)$ coincides for the first n digits leading to $\|f(t) - f(x)\|_2 < \sqrt{d} 2^{-n}$ which gives

$$\|f(t) - f(x)\|_2 < \sqrt{d}|t - x|^{\frac{\ln 2}{d \ln 3}}.$$

Choosing two points in Γ which differ exactly in the dn-th digit of their ternary representation proves conversely that the Hölder exponent cannot be larger on $\Gamma \subset I$. $\square$

The Hölder exponent changes if we change the construction of the Lebesgue curve using a different set than the Cantor set Γ where the expression $\log_2 3$ comes from. We can increase the factor. However, the limit factor of $1/d$ means that the set Γ approaches $\emptyset$ and the resulting curve is disconnected.

Now we have space-filling curves which are nowhere differentiable, but Hölder continuous with exponent $1/d$ like the Peano, Hilbert, Sierpiński and generalised Sierpiński curves, and we have the almost everywhere differentiable Lebesgue curve, which is of lower Hölder continuity $\frac{\ln 2}{d \ln 3}$ in the non-differentiable points. The question remains whether we can achieve higher Hölder continuity everywhere or even differentiability everywhere. The answer is negative, as we see in Lemma 4.12.

Lemma 4.12. *An everywhere differentiable curve with Lebesgue measurable image has an image of measure zero. It is not space-filling.*

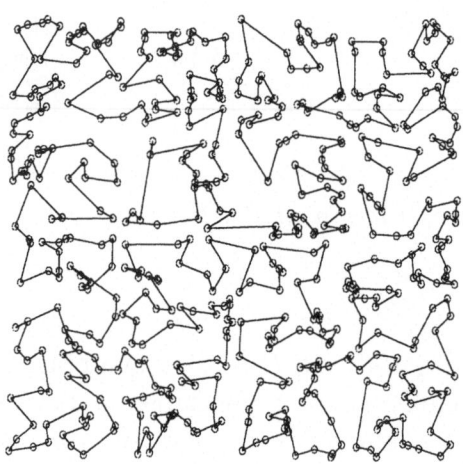

Figure 4.16. *A travelling salesman tour defined by a closed Hilbert-Moore space-filling curve.*

The proof of Lemma 4.12 is based on a theorem by Morayne [216], who considers the problem for functions $\mathbb{R} \mapsto \mathbb{R}^d$.

We will use space-filling curves for mesh partitioning. Here we will see that the Hölder continuity of $1/d$ is already optimal.

4.2   Partitioning

Space-filling curves had been created for purely mathematical purposes. However, nowadays there are a number of applications for space-filling curves. Basically, by the inverse of a discrete space-filling curve multi-dimensional data is mapped to a one-dimensional sequence. This mapping is useful for load balancing of parallel simulations on a computer, for data locality of memory or disc accesses inside a computer in geographic information systems [6, 75], for approximating shortest paths in optimisation [25, 38, 114, 123, 224], see Figure 4.16, ordering data in computer graphics [240, 298, 300, 303] and in other applications such as finding distances of point sets [32].

How can space-filling curves contribute to the efficient parallelisation of the introduced applications? The basic idea is to map the nodes or elements in space Ω to points on one iterate of a space-filling curve. The points can be mapped to the unit interval I by the inverse mapping of the space-filling. The points which now lie on the unit interval I can be sorted and grouped

```
int key2proc(keytype* separator, keytype k) {
  int numprocs;
  MPI_Comm_size(MPI_COMM_WORLD, &numprocs);
  for (int i=1; i<=numprocs; i++)
    if (k >= separator[i])
      return i-1;
  return -1; //   unreachable
}
```

Figure 4.17. *Compute the owner of a given node with key* k *according to the separators* s_j. *An improved imlementation can be based on binary search.*

together. The set of points can be partitioned on the interval I where each group is mapped to a processor. This defines a partition of the original problem in space Ω.

Partitioning by space-filling curves has been employed for finite element computations by Ou, Ranka and Fox [234] and by Oden, Patra and Feng [226] and has been compared to other heuristics by Pilkington and Baden [243]. The main advantage of space-filling curves in this context is their simplicity: if we want to bisect a set S of points $x_i \in \Omega$ on a space-filling curve that has been technically modified to be injective, we can do this by a single number s and the inverse space-filling curve mapping f^{-1}

$$S = \{x_i | f^{-1}(x_i) \leq s\} \cup \{x_i | f^{-1}(x_i) > s\}.$$

Each point is either left or right of the reference s on the space-filling curve. We take s as the median of $f^{-1}(S)$: both subsets are of the same size. In the same way we can partition the set of points S into p different subsets, if we partition $f^{-1}(S) \subset [0,1]$ by $p-1$ separators s_i like $\{x_i | s_j < f^{-1}(x_i) \leq s_{j+1}\}$. There is almost no bookkeeping necessary, because the partition is deterministic and can be computed on the fly from the separators. Each subset is assigned to one processor. Only the $p-1$ separators have to be stored, see Figure 4.35. A pseudo code example to compute the mapping of a node to the processor is shown in Figure fig:key2proc using MPI primitives, see [281].

We can solve the resulting one-dimensional partition problem: we cut the interval I into disjoint sub-intervals I_j of equal workload with $\bigcup_j I_j = I$. This gives perfect load balance and small separators between the partitions. The partition $f(I_j)$ of the domain Ω induced by the space-filling curve with $\bigcup_j f(I_j) \supset \Omega$ also gives perfect load balance. However, the separators $\partial f(I_j) \setminus \partial \Omega$ are larger than the optimal separators in general, as we will see. Geometric entities which are neighbours on the interval are also neighbours in the volume $\mathbb{R}^d$. Unfortunately the reverse cannot be true, and neighbours in the volume may be separated through the mapping.

There are different ways to implement the first step of the partitioning procedure, depending on the type of space-filling curve in use. We have to map a node or the location of some other geometric entity onto an iterate of a space-filling curve in space. We look for a point on the curve in the neighbourhood and compute its position on the space-filling curve.

For Lebesgue-type of space-filling curves this can be done easily: we round the coordinates, for example of the node (x, y) inside the unit square, to a finite binary representation of k digits. This results in a point which is on the kth iterate of the Lebesgue curve and all further iterates

$$\begin{pmatrix} x \\ y \end{pmatrix} \rightarrow \begin{pmatrix} 0_2.x_1x_2x_3 \dots x_k \\ 0_2.y_1y_2y_3 \dots y_k \end{pmatrix} \rightarrow 0_2.x_1y_1x_2y_2 \dots x_ky_k \ .$$

This procedure can be equivalently described as: compute the address of the quad-tree cell of the k-th level where the node resides. The d-dimensional case can be treated analogously.

Other type of space-filling curves require slightly more involved procedures to find a point on the curve next to the node and do the inverse mapping. However, the procedure related to the Hilbert curve can be described as a post-processing step of the Lebesgue procedure above. In a single loop from 1 to k the tuples $(x_i, y_i, \dots)$ can be transformed into tuples $(\hat{x}_i, \hat{y}_i, \dots)$ of the corresponding Hilbert curve, which is a similar process as the recursive construction of the Hilbert curve.

Further post-processing procedures can be used for the Sierpiński/ H-index curve and with a ternary number representation also for Peano curves. The generalised Sierpiński curve mapping can be implemented using local (barycentric) coordinates within each coarse element E of the initial mesh τ_0. Alternatively, the inverse mapping may be calculated following the mesh refinement rules according to Lemma 4.10.

We have to compute the separators such that the partitions contain the same number of nodes. This can be accomplished on a list of nodes sorted by their position on the space-filling curve $f^{-1}(x_i)$. The list is cut into equally-sized pieces and each piece is mapped to one processor. Hence the partition of nodes can be done with an ordinary sorting algorithm. Given a set of nodes x_i with their keys $f^{-1}(x_i)$, we first create a sorted list of keys. However, we need to perform the partition algorithm on a parallel computer. Given a set of nodes with their keys, which are distributed over the processors, we look for a parallel sort algorithm which results in a partition where each partition

```
void balance(keytype* separator) {
  sendNodes2Owner(separator);  // transfer nodes to processors

  int myrank, numprocs;
  MPI_Comm_size(MPI_COMM_WORLD, &numprocs);
  MPI_Comm_rank(MPI_COMM_WORLD, &myrank);
  long oldcount[numprocs], olddist[numprocs+1], newdist[numprocs+1];

  long c = countNodes();                    // amount of local work
  MPI_Allgather(&c, 1, MPI_LONG,
                &oldcount, numprocs, MPI_LONG, MPI_COMM_WORLD);
  olddist[0] = 0;
  for (int i=0; i<numprocs; i++)
    olddist[i+1] = olddist[i] + oldcount[i];// current distribution

  for (int i=0; i<=numprocs; i++)           // balanced distribution
    newdist[i] = (i * olddist[numprocs]) / numprocs;

  keytype newsep[numprocs+1];
  findSeparator(olddist[myrank], newdist, newsep);
  MPI_Allreduce(newsep, separator, numprocs+1, // new separators
                MPI_KEYTYPE, MPI_MAX, MPI_COMM_WORLD);

  sendNodes2Owner(separator);    // transfer nodes to processors
}
```

Figure 4.18. *Parallel partitioning by a bucket sort step. Send data to its respective owners according to separators s_j, balance the number of nodes per processor and adjust the separators s_j and send data to its respective owners according to the new separators $\hat{s}_j$.*

resides on the appropriate processor. There is no need to gather all keys on a master processor, but a pure scalable parallel algorithm performs better with respect to communication volume, memory usage and scalability. A pseudo code example is shown in Figure 4.18, see also Figure 4.35. A subroutine sendNodes2Owner is used to scan the nodes for ones, which do not belong to the processor according to the separators s_j stored in **separator**. These nodes are collected, sent to their respective owners and stored there. The countNodes function returns the number of nodes or some measure for the amount of work associated with the nodes. The findSeparator procedure scans the nodes in order to identify the nodes with numbers newdist, as long as they are on the processor. The results are combined by an MPI reduction operation, which gives the new separators $\hat{s}_j$. The subroutine sendNodes2Owner again transfers the nodes, where they belong to according to the new separators.

We present a single step bucket sort, where the previous separators serve as the bounds of the buckets. This is a good initial guess for the sorting procedure, provided that only small changes occur or that new load is balanced already. Hence for uniform mesh refinement e.g., only two steps of local neighbour communication are required, while in more difficult cases, two steps with more complicated communication patterns are necessary.

In addition to meshes, particles can also be partitioned by space-filling curves. Particle or n-body problems are defined by the interaction of n entities by some interaction forces. This model describes different phenomena, like the movement of planets or dust under gravitational forces in astrophysics and the dynamics of atoms or groups of atoms in molecular dynamics. The number of particles of interest easily reaches the range of $10^6 - 10^9$. The model leads to a system of n ordinary differential equations. The right hand side of each equation consists of the $n - 1$ forces of particles which interact with the particle under consideration. Usual model forces decay with the distance of the particles, which can be exploited by efficient approximation algorithms like the fast-multipole and the Barnes-Hut algorithm. However, there is still a global coupling of the particles which cannot be neglected. Furthermore, particles can be distributed randomly and can form clusters. Hence a parallel particle simulation code requires efficient load balancing strategies for the enormous amount of data. The particle move in space is due to the acceleration imposed by the interaction-forces. This means that a re-balancing or a dynamic load balancing is needed for parallel computing, which can be done by space-filling curves, see Warren and Salmon [302], Salmon, Warren and Winckelmans [262], and Parashar and Browne [237].

A slightly different situation can be found in adaptive discretisations of partial differential equations. Now a mesh consisting of n nodes and elements or volumes has to be distributed to a parallel computer. The nodes and elements can be found at arbitrary positions (completely unstructured meshes) or at fixed positions which are a priori known (structured meshes) or at least computable by mesh-refinement rules from a coarse mesh (adapted meshes). The degrees of freedom are coupled locally, usually between neighbouring nodes. However, algorithms for the solution of the resulting equation systems like direct methods or multilevel methods couple all degrees of freedom together, which imposes difficulties on the parallelisation of the solver. For the solution of stationary problems, no nodes are moved in general. However, due to adaptive mesh refinement, new nodes are created during the computation. This requires dynamic load balancing, which can also be done by space-filling curves [90, 151, 226, 237, 256, 271].

The partitioning problem in general is NP-hard, see Bui and Jones [76]. There are many heuristics based on graph connectivity or geometric properties to address this problem [27, 100, 113, 183, 245]. In practice, fast heuristics are known. However, not much is known about the general quality of these methods. On the very contrary there are examples where single heuristics fail and give bad results.

We use a basic performance model for a parallel computer. The execution time of a program consists of computing time, which is proportional to the number of operations on a processor, and of communication time. Communication between the processors consumes time proportional to the size of data.

We consider fast $\mathcal{O}(n)$ algorithms linear in the size of data n, e.g. FEM matrix assembly, sparse matrix multiply or components of a multigrid algorithm such as a grid transfer or smoother. The parallel computing time is $C_1 \cdot n/p$ for a partition of n data onto p processors with C_1 some constant time per mesh point. We define $v := n/p$ as the *volume*. The runtime depends on the communication time which we assume to be linear in the amount of data to be transferred. It is proportional to the separator or surface s_j of the partition $s_j := \partial f^{-1}(I_j) \setminus \partial \Omega$ with C_2 some constant time per mesh point

$$t_p = C_1 \frac{n}{p} + C_2 s_j \,. \tag{4.4}$$

This model suggest that we have to minimise the surface to volume ratio s_j/v for each partition in order to achieve good load balancing. Looking at a generic partition with $s = \max_j s_j$ we want to achieve a high parallel efficiency of

$$\text{efficiency} \;=\; \frac{t_1}{p \cdot t_p} \;=\; 1/(1 + \frac{C_2}{C_1} \frac{s}{v}) \,. \tag{4.5}$$

The lowest ratio of surface to volume s/v for continuous manifolds is obtained for the sphere by

$$s = \sqrt[d]{2d^{d-1} \frac{\pi^{d/2}}{\Gamma(d/2)}} \; v^{(d-1)/d} \,.$$

However, we usually deal with partitions aligned with the mesh. Hence the cube with

$$s = 2d v^{(d-1)/d} \tag{4.6}$$

is of interest. In general we regard estimates of type

$$s \leq C_{\text{part}} \cdot v^{(d-1)/d} \tag{4.7}$$

with low constants C_{part} as optimal.

The discrete counterpart of the surface to volume ratio is given by the isoperimetric number. It is defined by the maximum ratio of v/s over all partitions of fixed size v of a given mesh. In graph notation this is $\max_A \frac{\min(|A|,|B|)}{\partial A}$ for a graph with n vertices and all possible partitions $A \cup B$ of the graph into disjoint sets A and B. It is known that isoperimetric number is of the order $\mathcal{O}(n^{1/d})$ for geometric meshes in $\mathbb{R}^d$, which is the analog to Equation (4.7). However, we are interested in a bisection of the graph with $|A| = |B|$ or $|A| = |B| + 1$ rather than arbitrary partition sizes.

There are some results for the quality of partitions obtained by graph partitioning methods, which we will summarise briefly. While it is known that the spectral bisection scheme based on the Fiedler vector of the discrete Laplacian due to Pothen, Simon and Liou [245] may fail to give cut sizes smaller than $\mathcal{O}(n)$ for an n node graph, see Guattery and Miller [157], the spectral bisection with best-ratio cut scheme can be iterated. In this way the separator of the graph is always of the quasi-optimal size for the smaller partition, and the larger one is recursively bisected for well shaped graphs. The underlying geometrical meshes must be standard FEM meshes with elements of bounded aspect ratio and bounded interior angles, see Spielman and Teng [284]. More precisely, the mesh fulfils an α-overlap condition where each vertex of the graph is the centre of a sphere S_i in $\mathbb{R}^d$. The spheres are all disjoint $S_i \cap S_j = \emptyset$ for $i \neq j$. However, an edge (i, j) in the graph is equivalent to the condition

$$\alpha S_i \cap S_j \neq \emptyset \quad \text{and} \quad S_i \cap \alpha S_j \neq \emptyset$$

where αS_i denotes the sphere S_i with α times the radius. Shape-regular finite element meshes are α-overlap meshes, as are many other graphs.

As a result, the iterated spectral bisection scheme gives a separator of quasi-optimal size $\mathcal{O}(n^{(d-1)/d})$. However, the graph is not partitioned into equal sized parts, but the smaller partition contains at least $n\frac{1}{d+2}$ vertices. The same partition estimates have been derived for a geometric bisection scheme by Miller, Teng, Thurston and Vavasis [212]. The vertices are bisected by a sphere in $\mathbb{R}^d$. This sphere is constructed algorithmically by a spherical projection of data onto a hyper-sphere in $\mathbb{R}^{d+1}$, the construction of a centre-point in $\mathbb{R}^{d+1}$ and a hyper-plane cut. The centre-point has to be constructed by a heuristic for complexity reasons. Furthermore, the hyper-plane is also selected by a randomised algorithm, so that a good partition is found with high probability. This modified scheme will be faster than the iterated spectral bisection. Both methods can be parallelised. A modification of the geometric bisection scheme for balanced bisections is possible, but the cut size estimate is lost as for the original spectral bisection method.

A similar construction with centre-point and cutting plane was proposed in $\mathbb{R}^d$ instead of $\mathbb{R}^{d+1}$. The projection onto a sphere is omitted and a plane cut is used instead of a sphere cut. For quasi-uniform, well shaped meshes and the ratio k of longest to shortest edge in the whole mesh, a separator of size $\mathcal{O}((a + \log_2 k)^{1/d} n^{(d-1)/d})$ and a balance of $n^{\frac{1}{d}}$ was proven by Cao, Gilbert and Teng [83]. Balanced bisections with $1 : 1$ partition sizes can be obtained for slightly larger separators of size $\mathcal{O}((1 + \log_2 k)^{1/d} n^{(d-1)/d} \log_2^{1/d} n)$. Note that the factor k can be substantial. For adaptively refined meshes $\log_2 k$ is proportional to the number of mesh levels, and we may obtain only sub-optimal results on separator sizes.

So far we have discussed partitions into two parts. A graph can be partitioned into more parts by recursively applying a bisection scheme. There are graphs where even a perfect bisection scheme fails to give good partitions into more parts, see Simon and Teng [275]. However, for well shaped meshes like α-overlap meshes, which have a property that they have families of $f(n)$ bisectors with a sub-linear function f, the recursive bisection works well. Especially for a $\mathcal{O}(n^{(d-1)/d})$ bisection scheme and well shaped meshes, the p-way partition gives a cut size of $\mathcal{O}((n/p)^{(d-1)/d})$. Note that the partitions will not be balanced well.

Space-filling curves, in contrast, provide a method to partition a mesh into an arbitrary number of partitions, each of arbitrary size. Hence we can construct balanced p-way partitions and we will derive quasi-optimal separator sizes $\mathcal{O}(n^{(d-1)/d})$ also in this case, where the analysis of other graph partitioning methods fails so far. However, it is not known whether any mesh can be partitioned with quasi-optimal separator and balanced partitions and we will describe additional criteria for such meshes.

First of all, we do have the right tools so far to prove optimality for quasi-uniform meshes, i.e. bounded variation of element volumes of a mesh. We consider the continuous case of space-filling curves where we have Hölder continuity for certain curves. Bounds of the surface of partitions of type Equation (4.7) will be derived in the following lemma.

Lemma 4.13. *Given a self-similar $1/d$-Hölder continuous space-filling curve $f : I \mapsto \Omega$ parametrised by volume, the surface s of a partition $f([t, t + v))$ of size v is bounded by Equation (4.7). The constant C_{part} depends on the constant C of Hölder continuity (Equation (4.2)).*

The proof is based on Equation (4.2). An axis parallel box of side length $2C \sqrt[d]{v}$ and centre $f(t)$ in Ω does enclose the image $f([t, t+v))$ due to Equation

(4.2). Its surface is $s_b = 2d(2C\sqrt[d]{v})^{d-1} = 2^d C^{d-1} v^{(d-1)/d}$ (Equation (4.6)). Furthermore, self-similarity and the construction scheme by a template guarantees that the surface of the set $f([t, t+v))$ is not fractally shaped and especially that it can be bounded by a multiple of s_b. □

The proof is not applicable to Lebesgue-type curves, also called *bit interleaving* by Beichl and Sullivan [32], because its Hölder continuity is not sufficient and the discrete partitions tend to be disconnected. However, Lemma 4.13 also holds for a Lebesgue curve, which is shown directly in two dimensions in [180]. The constant C_{part} derived in Lemma 4.13 is only an upper estimate. Sharper bounds can be obtained by directly looking into the proof of Hölder continuity of the respective space-filling curve, see also the remarks on proof of Lemma 4.3.

This setting can be carried over to the discrete case of finite size elements if the curve is parametrised by volume: an element E of volume $1/n$ is the image of an interval I_e of length $1/n$, $\overline{f(I_e)} = \overline{E}$. The interval I is cut into n pieces.

We can relax this further for the case of quasi-uniform meshes.

Lemma 4.14. *We assume a self-similar $1/d$-Hölder continuous space-filling curve $f : I \mapsto \Omega$ which is parametrised by volume as in Lemma 4.13. Further we assume a quasi-uniform mesh. An estimate of type Equation (4.7) holds for a partition with v the number of elements and s the number of elements faces.*

The approach can be interpreted as a re-parametrisation of the curve *by element*.

The proof is based on Lemma 4.13 and the quasi-uniformity of the mesh τ. Each element $E \in \tau$ of a quasi-uniform mesh τ has a volume v_e between $v_{\max}/c_{\min} < v_e \leq v_{\max}$ for a maximum element volume $v_{\max}$. Furthermore the area of an element face s_e is bounded because the element volumes are bounded and angle and vertex conditions hold, $0 < s_{\min} \leq s_e \leq s_{\max}$. The continuous estimate $s \leq C_{\text{part}} v^{(d-1)/d}$ translates into a discrete one with a different constant:

$$s_{\text{discr}} \leq \frac{C_{\text{part}}}{s_{\min}} \left(\frac{v_{\text{discr}}}{v_{\max}}\right)^{(d-1)/d}.$$

□

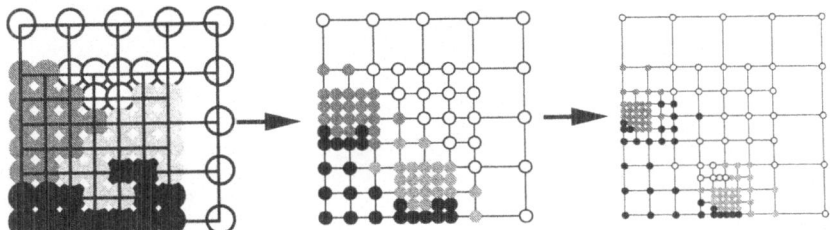

Figure 4.19. *A sequence of adaptively refined 2D meshes mapped to four processors, partition is grey coded.*

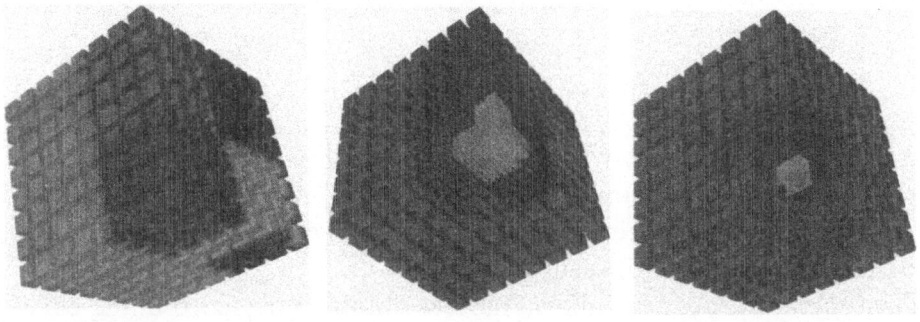

Figure 4.20. *A sequence of adaptively refined 3D meshes mapped to 8 processors, partition is grey coded.*

Corollary 4.15. *Estimate (4.7) also holds for a space-filling curve partitioning of a (quasi-) uniform mesh covering $\tilde{\Omega} \subset \Omega$ by superposition of $f : I \mapsto \Omega$ or mesh dependent construction of $f : I \mapsto \tilde{\Omega}$.*

Lemma 4.14 combined with Equation (4.5) gives a parallel efficiency of at least

$$\text{efficiency} = 1 \Big/ \Big(1 + \frac{C_2 C_{\text{part}}}{C_1} \cdot \sqrt[d]{\frac{p}{n}}\Big) . \tag{4.8}$$

This implies optimal parallel efficiency for very large problems, $n \to \infty$. Estimate (4.8) holds for a code for the solution of partial differential equations in the steps of setting up an equation system, a single matrix multiply, a fixed number of Krylov iterations. Furthermore, using the same space-filling curve at all mesh levels, this also holds for an additive multigrid implementation and for standard multigrid if we neglect effects of global communication and

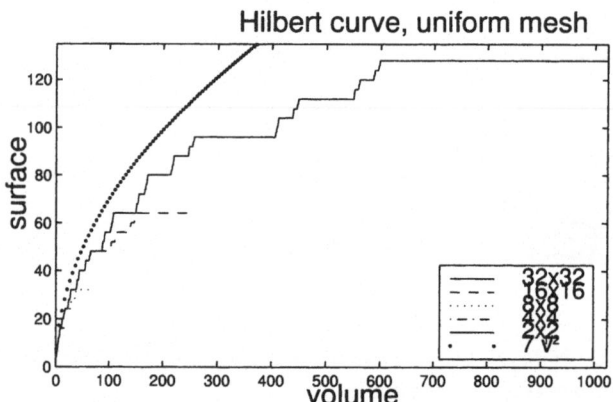

Figure 4.21. *Locality of partitions defined by a space-filling curve. Hilbert curve on a uniform mesh.*

number of levels, as we do with the simple computer model (Equation (4.4)). For the scalability of a global PDE solver an $O(n)$ multigrid solver is essential. Solvers with higher than linear complexity may scale in p like Equation (4.8) but do scale completely different in n.

The separator sizes for uniform meshes obtained by space-filling curve partitioning are optimal up to a multiplicative constant. This indicates that the space-filing curve partitioning algorithm performs asymptotically well in this case. The proof of Lemma 4.14 only gives a crude estimate on the constant in Equation (4.7). Hence we look at two examples for two-dimensional partitions. In Figure 4.21 the maximum surfaces s to different volumes v are given. We consider a uniform square mesh $[0, 2^k]^2 \subset \mathbb{Z}^2$ (counting the complete boundary) and a triangulation (counting the interior boundary only). The triangulation starts with a hexagon and angles of $\pi/3$ and is refined adaptively. The triangulation is shown in Figure 4.23 left. The different graphs in Figure 4.21 show the ratios for different mesh levels. The surface of small partitions comes close to the expected $\sqrt{n}$ behaviour while larger partitions have a boundary which is limited by the boundary of the domain $\partial\Omega$.

4.3 Partitions of Adaptively Refined Meshes

The Lemmata 4.10 and 4.14 do not deal with adaptive mesh refinement. We want to extend the space-filling curve partitioning heuristic to adaptively refined meshes. What happens if we apply the heuristic? On a fixed level an adaptive mesh can always be interpreted as a quasi-uniform mesh, where the

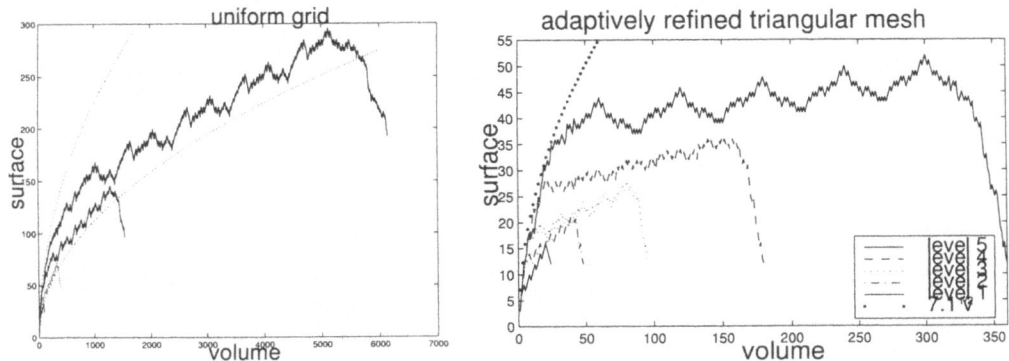

Figure 4.22. *Locality of partitions defined by a two dimensional generalised Sierpiński space-filling curve defined on an unstructured mesh with uniform (left) and adaptive mesh refinement (right), compare Figure 4.23 left and middle.*

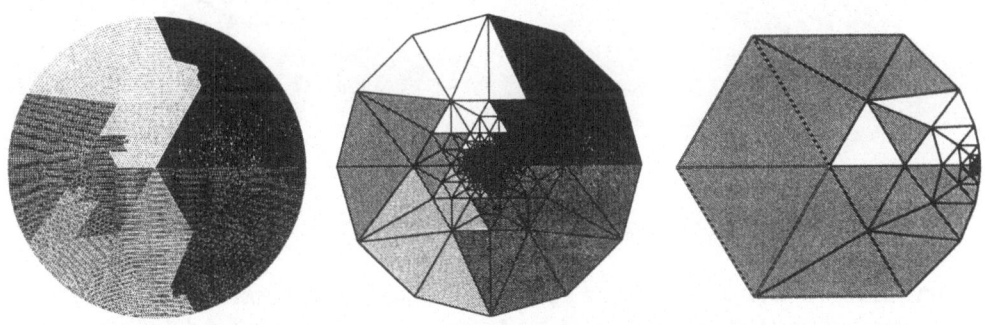

Figure 4.23. *Adaptive mesh refinement. Partitions defined by a space-filling curve. Uniform nesh refinement (left, Lemma 4.14), adaptive (middle, Definition 4.16) and steep adaptive refinement (right, Example 1).*

method performs well. However, with increasing level of mesh refinement the constant $c_{\min}$ decreases and the estimate on the partitions C_{part} in Equation (4.7) gets worse. In the limit case, the parametrisation by element does not even lead to a (space-filling) curve.

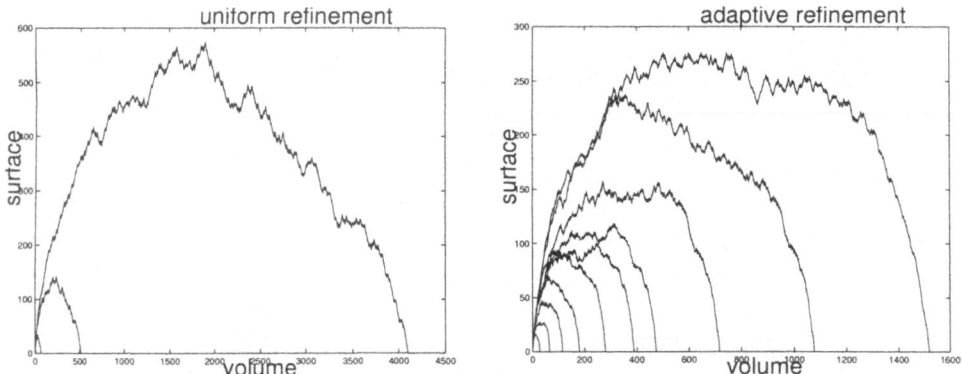

Figure 4.24. *Locality of partitions defined by a three dimensional generalised Sierpiński space-filling curve defined on an unstructured mesh with uniform (left) and adaptive mesh refinement (right), compare Figure 4.25 left and right.*

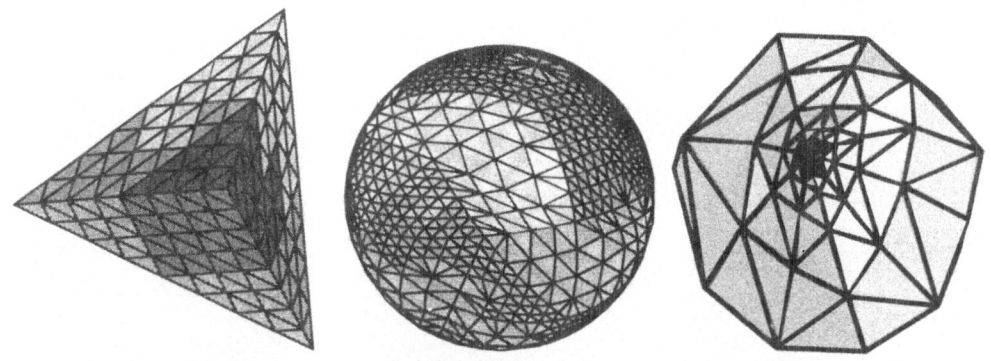

Figure 4.25. *Partitions defined by a generalised Sierpiński space-filling curve. Uniform meshes (left and middle) and adaptive refinement (right).*

4.3.1 Arithmetic Progression of Nodes

Example 1. *Let us consider a case of very steep adaptive mesh refinement with arithmetic progression of the element number. An initial mesh τ_0 of a domain Ω is refined in each refinement step in a way that the number of elements increases by an additive constant c, $|\tau_i| \leq |\tau_0| + ci$. Only the elements adjacent to a single point $a \in \partial\Omega$ on the boundary are subdivided plus some further elements necessary due to mesh refinement rules, see Figures 4.26, 4.27, 4.28 and 4.23 (right). Now the sequence of curves defined by the sequence of meshes does not converge to a space-filling curve. The obvious reason is that they do*

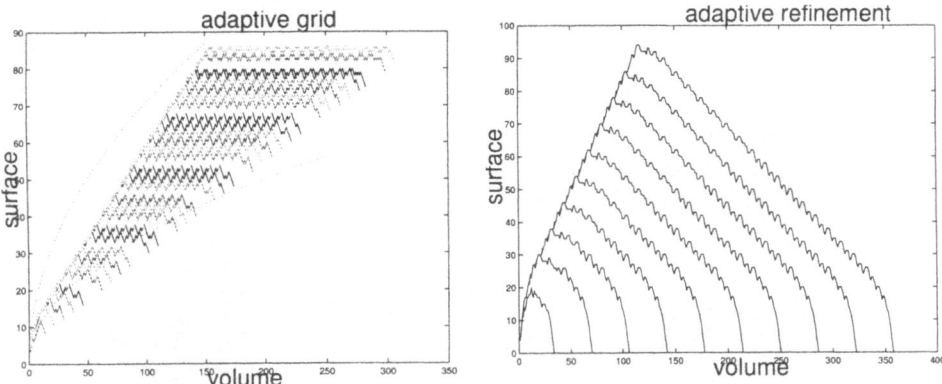

Figure 4.26. *Generalised Sierpiński-type space-filling curves on meshes with steep adaptive mesh refinement of arithmetic progression as in Example 1. 2D (left) and 3D (right), compare Figure 4.23 right and Figure 4.28 left.*

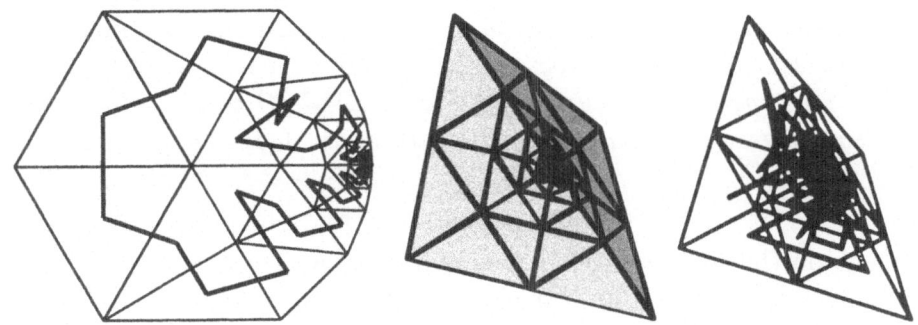

Figure 4.27. *Generalised Sierpiński-type space-filling curves on meshes with steep adaptive mesh refinement of arithmetic progression as in Example 1.*

not fill the domain except for a vicinity of the point a. However, we can change the construction by occasionally inserting a uniform mesh refinement step into the refinement process. If we insert the i-th uniform refinement step, which increases the number of elements by a factor of k, after (k^i) adaptive steps, the number of mesh nodes will still increase arithmetically.

In Example 1 we have a quasi-uniform refinable sequence of meshes with an arithmetic progression of elements at hand. Elements next to the point a are refined in every refinement step. A space-filling curve parametrised by volume can be defined, see Lemma 4.10. However, the curves parametrised

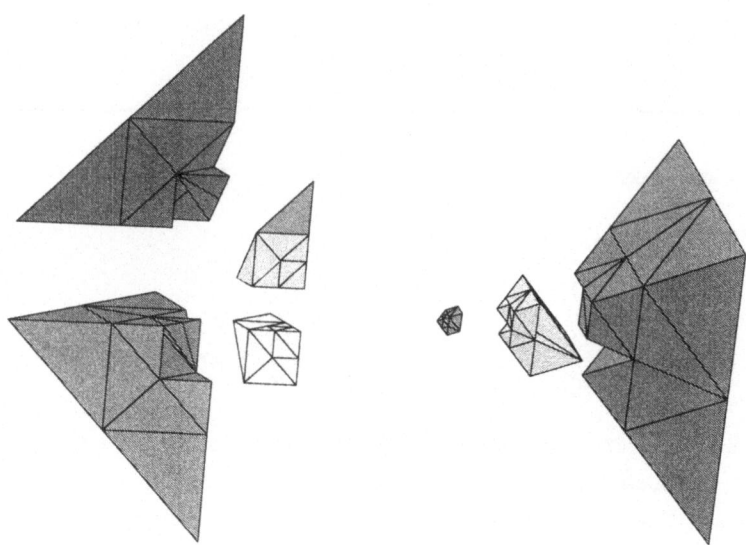

Figure 4.28. *Adaptive mesh refinement. Four partitions defined by the generalised Sierpiński space-filling curve for steep adaptively refined tetrahedral meshes with arithmetic progression. See Example 1 and the generalised Sierpiński-type space-filling curve of Figure 4.27 (left) and a similar configuration also according to Example 1 with minimum surface (right).*

by element on the meshes τ_i do not converge to a continuous curve, because a decreasing and in the limit vanishing part of the interval I is mapped to any part of Ω which is a distance $\epsilon > 0$ away from a. Hence an estimate for discrete partitions similar to Lemma 4.14 is not possible this way. Looking at numerical examples for this scenario, Figure 4.26, we see that estimates of type Equation (4.7) indeed do not hold. Rather we obtain a worst case scenario of

$$s = \mathcal{O}(v) \,,$$

where s is proportional to v and a parallel efficiency that is constant independent of the problem size n. This behaviour limits the usefulness of the partition method for this special case. The two dimensional *Counter* Example 1 is related to the roach graph proposed by Guattery and Miller [157] where standard graph partitioning heuristics like spectral bisection [245] gives similar partitions and fails to perform well.

Note that the situation will look different if we are able to change or modify the space-filling curve. Moving the singularity to one end of the space-filling curve, the resulting partitions provide the necessary locality again. This can be shown using a re-parametrisation of the curve by some diffeomorphism, e.g. $x \mapsto x^{\alpha}$.

In the following we will argue that the case of arithmetic progression in the number of elements does not occur for the solution of elliptic boundary value problems. Furthermore, mesh partitioning in such cases performs very well, often comparable to the case of uniform mesh refinement as it is illustrated also in Figures 4.22 and 4.24. The case of best n-term approximation is covered later in this section, where we indeed obtain geometrically graded meshes.

4.3.2 β-adaptive Mesh Refinement

Definition 4.16. *We define an β-adaptive family T of meshes as a quasi-uniform refinable family of meshes with an additional condition. Each element $e_{i,j}$ of a mesh τ_i is subdivided into at least two smaller elements in mesh τ_{i+1} reflected by the constants $2 \le C_2 \le C_1 < \infty$*

$$C_2 \le |\{e_{i+1,k} \in \tau_{i+1} | e_{i+1,k} \subset e_{i,j}\}| \le C_1 .$$

The value $\beta := \log C_1 / \log C_2 \ge 1$ is a measure for the differences of the number of refinement steps taken in different areas of the domain Ω.

Many meshes used for the solution of partial differential equations are in fact β-adaptive with some value β. Of course, sequences of uniform and quasi-uniform meshes fulfil the conditions. Furthermore, adaptive mesh refinement procedures may lead to β-adaptive meshes if the mesh refinement is controlled in a certain way. However, the mesh constructed in Example 1 is not β-adaptive, because the value $\beta \to \infty$ is not bounded for increasing refinement steps.

We argue that the approximation of a solution u of an elliptic boundary value problem by means of a 'standard' discretisation scheme automatically leads to β-adaptive meshes with β a measure of the variation of local regularity of the solution u. Given a boundary value problem

$$\mathcal{L}(u) = f \text{ in } \Omega \qquad \text{and } g(u) = 0 \text{ on } \partial\Omega$$

with an elliptic second order differential operator $\mathcal{L}$ with smooth coefficients, a sufficiently smooth function f, a Lipschitz domain Ω, we assume that some suitable boundary conditions are encoded in the operator g. The solution u can be decomposed into several components. In the interior of the domain Ω the solution u is smooth. Singularities occur due to boundary conditions and

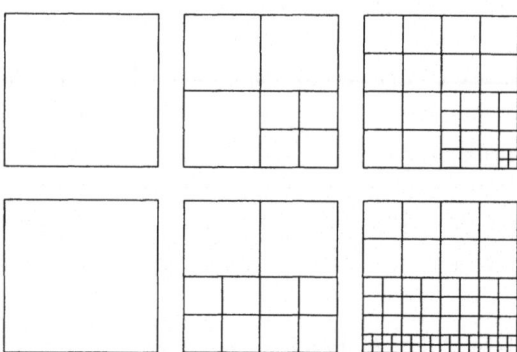

Figure 4.29. *β-adaptive mesh refinement for a point-singularity (top row) and an edge-singularity (bottom row).*

the geometric shape of the boundary $\partial\Omega$. Typical types of singularities are vertex singularities located at a cone point of $\partial\Omega$ where u is asymptotically r^α in spherical coordinates and for $d > 2$ edge singularities located at an edge of $\partial\Omega$ where u is asymptotically r^α in cylindrical coordinates. If we subtract the singular components from u, the remaining part is smooth. Now the convergence rates in the energy norm are locally of the form $\mathcal{O}(h^\alpha)$ with $\alpha = 1$ for smooth parts of the solution and lower α next to a singularity. Given a mesh τ such that the error in the energy norm is ϵ and is roughly equidistributed over the domain Ω, a reduction of the error to the value $\epsilon/2$ requires mesh refinement. In an area with local convergence $\mathcal{O}(h^\alpha)$, the characteristic mesh size h has to be reduced by a factor of $2^{-1/\alpha}$. Using isotropic element refinement, this leads to a local increase in the number of elements by a factor of $2^{-d/\alpha}$. Given the minimum and maximum values of α, we obtain a β-adaptive mesh with

$$\beta = \alpha_{\max}/\alpha_{\min}\,.$$

Lemma 4.17. *Let $f : I \mapsto \Omega$ be a self-similar $1/d$-Hölder continuous space-filling curve parametrised by volume as in Lemma 4.13. Then on a β-adaptive family of meshes an estimate*

$$s \;\leq\; C_{\mathrm{part}}C_1 \cdot v^{\beta\frac{d-1}{d}}$$

holds for a partition with v the number of elements and s the number of element faces.

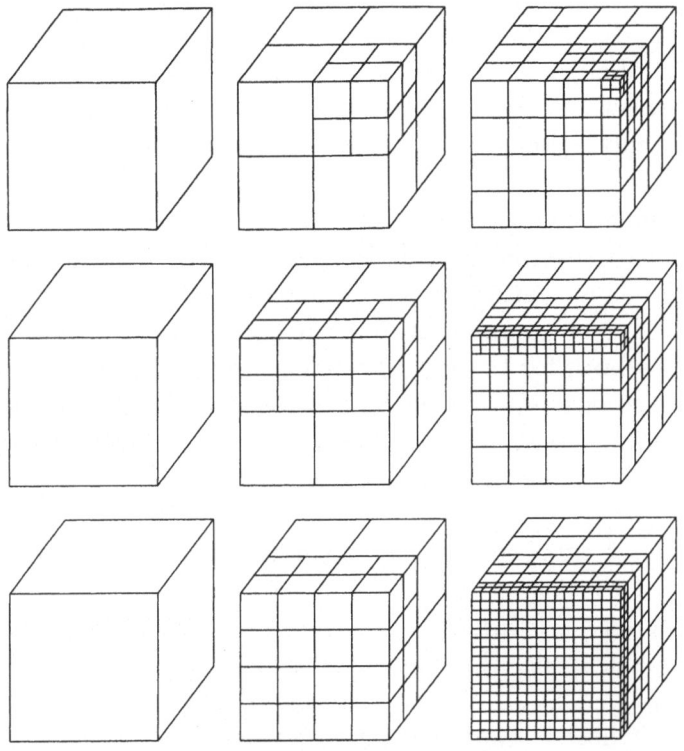

Figure 4.30. *β-adaptive mesh refinement for a point-singularity (top row), an edge-singularity (middle row) and a face-singularity (bottom row).*

Unfortunately the proof cannot be based on Lemma 4.13. Note that a re-parametrisation *by element* of the space-filling curve is not possible for $\beta > 1$, because the fraction of the number of elements located in one area is in the limit either zero or one. Hence we show an analogon to property (d), the Hölder continuity of Lemma 4.10 of the discrete curve.

Given a mesh τ_i of n_i elements and two arbitrary elements e_1 and e_2. On the discrete generalised Sierpiński curve S_i through mesh τ_i the elements are v elements apart. Now we look for a coarser mesh τ_j with $j \leq i$ that is the finest mesh which fulfils the following condition. Element e_1 is part of an element $\hat{e}_1 \in \tau_j$ that contains at least v elements of mesh τ_i and also all neighbour elements of e_1 contain at least v elements of mesh τ_i, $e_1 \subset \hat{e}_1$. The curve S_i can be projected to the coarser mesh τ_j. The element e_2 is either also contained in $\hat{e}_1$ or it is in one of $\hat{e}_1$'s neighbour elements $\hat{e}_2$, because the projected curve

S_i first fills up a whole element before entering a neighbour element on mesh τ_j.

The number of elements of mesh τ_i which build up the elements $\hat{e}_1$ and $\hat{e}_2$ is between C_1^{i-j} and C_2^{i-j} that is below $C_1 v^\beta$, because the number of coarsening steps $i-j$ is at most $\lfloor \log v / \log C_2 \rfloor$. It remains to compute the surface of the complex $\hat{e}_1 \cup \hat{e}_2$.

Let us assume an element complex which is created by uniform subdivision of a single element several times subsequently. Due to the angle and vertex conditions, there is a relation of type Equation (4.7) between the surface and volume of the element complex in units of number of elements v and number of surface faces s both for the initial element and for all its refined complexes. The surface of the elements which build up both $\hat{e}_1$ and $\hat{e}_2$ is at most CC_2^{i-j}. Inserting the volume estimate v^β into this, the lemma is proven, with C_{part} dependent on the element angle and vertex conditions. $\square$

The estimate of this Lemma 4.17 includes the uniform and quasi-uniform case of Lemma 4.14 with $\beta = 1$. It is useful for values of β up to $\frac{d}{d-1}$. For larger values the trivial estimate $s \leq C_v v$ in units of number of surface faces s and elements v is sharper. However, this trivial estimate does not lead to any meaningful characterisation of a partition derived by the space-filling curve.

Our approach now is twofold. First, we construct meshes which demonstrate that the estimate is sharp for large $\beta = \frac{d}{d-1}$, because we know it is sharp for $\beta = 1$. Second, we give a further refined criterion to obtain tighter estimates of the type of the estimates for uniform meshes Lemma 4.14, but this time for adaptively refined meshes.

Example 2. *Let us consider a case of adaptive mesh refinement toward an area of interest S of codimension one at the boundary $S \subset \partial\Omega$, i.e. an edge for $d = 2$ or a face for $d = 3$. We construct a β-adaptive sequence of meshes with $\beta = 2$. Starting with some initial mesh τ_0, we first apply a uniform refinement step, followed by a refinement of all elements adjacent to S.*

Two sequences of meshes of this type can be seen in Figures 4.29 (bottom row) and 4.30 (bottom row). Let us assume that an element is subdivided into C^d sub-elements at each refinement step, the surface of the mesh at S increases roughly by a factor of C^{d-1} by a single refinement step and by a factor of $C^{2(d-1)}$ from mesh τ_i to τ_{i+1}. Hence the number of surface faces of the whole mesh τ_i increases in i as $\mathcal{O}(C^{2i(d-1)})$. The number of elements in the interior of the domain does increase only as $\mathcal{O}(C^{id})$, whereas next to the area S we obtain a growth rate of the order of the number of surface faces, $\mathcal{O}(C^{2i(d-1)})$, which is larger than the interior domain growth for $d > 2$. The ratio v/s is bounded for $d > 2$ and grows linearly in i for $d = 2$. The surface to volume ratio is far away from the optimal $s = \mathcal{O}(v^{(d-1)/d})$. This is in agreement with Lemma 4.17, because $\beta \geq \frac{d}{d-1}$. We conclude for Example 2 with the estimates

$$
s = \begin{cases} \mathcal{O}\left(\frac{v}{\#\text{levels } i}\right) & \text{for} \quad d = 2 \\ \mathcal{O}(v) & \text{for} \quad d > 2 \end{cases} .
$$

Note that the locality of the partitions can be improved if we change the curve. *Anisotropic* curves can be employed which give optimal estimates. They will be discussed later.

4.3.3 γ-adaptive Mesh Refinement

Under some sharper assumptions than β-adaptivity, where we restrict the number of elements by subdivision, we can give a tighter estimate than Lemma 4.17. In order to do this, we generalise the Definition 4.16 of β-adaptivity.

Definition 4.18. *We define the function $\gamma \geq 0$ for a quasi-uniform refinable family of meshes by the following condition. Each element $e_{i,j}$ of the mesh τ_i is subdivided into two or more smaller elements in mesh τ_{i+1} reflected by the mesh independent bounds $\underline{\gamma}(1)$ and $\overline{\gamma}(1)$ with $2 \leq \underline{\gamma}(1) \leq \overline{\gamma}(1) < \infty$. We assume that uniform mesh refinement is associated to the lower bound. Furthermore, this number of elements is refined uniformly in any case, so that only up to $\overline{\gamma}(1) - \underline{\gamma}(1)$ elements are created additionally by another procedure.*

The total number of elements created in refinement step k on mesh τ_{i+k} is a number of elements between the mesh independent bounds $\underline{\gamma}(k)$ and $\overline{\gamma}(k)$ with $2 \leq \underline{\gamma}(k) \leq \overline{\gamma}(k) < \infty$ with monotonically increasing γ as a function of k. Again, the uniform refinement part is at least $\underline{\gamma}(k)\overline{\gamma}(k-1)/\underline{\gamma}(k-1)$ with an additional part of elements created by another procedure.

For technical reasons, we assume that $\underline{\gamma}(k)$ increases exponentially, i.e. $\underline{\gamma}(k) = C_\gamma^k$ with $C_\gamma > 1$. This can be achieved by a re-parametrisation of the functions γ and the meshes τ, if a single set of uniform mesh refinement rules is used.

The definition does not restrict the set of quasi-uniform refinable meshes like β-adaptivity. However, we are interested in families of meshes with small ratios $\overline{\gamma}/\underline{\gamma}$. For example for β-adaptive meshes we immediately have the constant functions $\underline{\gamma}(i) = C_2^i$ and $\overline{\gamma}(i) = C_1^i$. Conversely, a γ-adaptive mesh will also be β-adaptive if $\beta := \overline{\lim}_{i\to\infty} \frac{\log \overline{\gamma}(i)}{\log \underline{\gamma}(i)}$ is finite, i.e. the ratio $\overline{\gamma}(i)/\underline{\gamma}(i)$ is bounded independent of i.

In more generality, the number of elements created by uniform refinement can be at least some function C^k with $C < C_\gamma$. In such a case, the following estimates will be weaker by the exponent $\log C_\gamma / \log C$, analogously to the estimates on β-adaptive meshes with $\beta > 1$.

Lemma 4.19. *Let $f : I \mapsto \Omega$ be a self-similar $1/d$-Hölder continuous space-filling curve parametrised by volume, on a γ-adaptive family of meshes with*

$$\sum_{j=0}^{\infty} \frac{\overline{\gamma}(j) - C_\gamma \overline{\gamma}(j-1)}{C_\gamma^{\frac{j(d-1)}{d}}} < \infty .$$

Then the estimate

$$s \leq C_{\mathrm{part}} \cdot v^{\frac{d-1}{d}}$$

holds for a partition with v the number of elements and s the number of element faces. However, if the sum grows linearly in j, we obtain the weaker estimate of

$$s \leq C_{\mathrm{part}} \cdot v^{\frac{d-1}{d}} \log v .$$

The proof is a refinement of the proof of Lemma 4.17. Again, we have to show a discrete analogon of Hölder continuity. Given a mesh τ_i of n_i elements and two arbitrary elements e_1 and e_2 and a distance v, we find a coarser mesh τ_j in at most $\log v / \log C_\gamma$ coarsening steps and two neighbouring elements $\hat{e}_1$ and $\hat{e}_2$ on the mesh such that the whole section of length v of the generalised Sierpiński curve resides within $\hat{e}_1 \cup \hat{e}_2$.

The surface of the complex $\hat{e}_1 \cup \hat{e}_2$ can be bounded sharper than in Lemma 4.17 based on the function $\overline{\gamma}$. The number of faces of any single element is bounded by C_3. From step $k-1$ to the next k, the number of elements increases by a factor between C_γ and $\overline{\gamma}(k)/\overline{\gamma}(k-1)$. At least the factor C_γ is due to uniform refinement, which maintains the appropriate surface to volume relation. The surface increases for a uniform refinement step by a factor of $C_\gamma^{(d-1)/d}$. However, with $\overline{\gamma}(k)/\overline{\gamma}(k-1) > C_\gamma$, additional elements and faces are created, such that we add at most $C_3 \overline{\gamma}(k)/\overline{\gamma}(k-1) - C_\gamma$ faces. Iterating this we get the upper bound

$$s \leq 2C_3 \left((C_\gamma^{(d-1)/d})^{i-j} + \sum_{k=j}^{i-1} \left(\frac{\overline{\gamma}(k-j+1)}{\overline{\gamma}(k-j)} - C_\gamma \right) \left(C_\gamma^{(d-1)/d} \right)^{i-1-k} \right) .$$

Together with $v \geq C_\gamma^{i-j}$ we get the final estimates. $\square$

It was possible to prove an estimate of type Equation (4.7) with exponent $(d-1)/d$ for the quasi-uniform case, but the β-adaptive case gives slightly weaker estimates of $\beta(d-1)/d$. For the γ-adaptive case we now have a criterion to obtain estimates of type Equation (4.7) with exponent $(d-1)/d$ also for adaptively refined meshes. We will discuss some typical situations of singularities and their related functions $\overline{\gamma}$. Now the strength α of an asymptotic r^α-expansion does not play the prominent role, but rather the relative area of mesh refinement. In Figures 4.29 and 4.30 some types are depicted. Note that the case of codimension one mesh refinement has been treated in Example 2 already with a sharp result that the surface is proportional $(d > 2)$ or only logarithmically $(d = 2)$ lower than the volume.

Example 3. *We construct a quasi-uniform refinable sequence of meshes. Starting with an arbitrary initial mesh τ_0, each mesh is created by a uniform refinement step and an adaptive refinement step. The mesh parameter h is divided by two at the uniform refinement step and the minimal mesh parameter is divided by two again at the adaptive refinement step. Elements adjacent to a g-dimensional sub-manifold S of Ω are refined during the adaptive refinement step, see also Figures 4.29 and 4.30.*

We obtain a γ-adaptive mesh with $C_\gamma = 2^d$ and a function $\gamma(j) = \mathcal{O}(4^{gj})$. The criterion of Lemma 4.19 reads

$$\sum_{j=0}^{\infty} \frac{\overline{\gamma}(j) - C_\gamma \overline{\gamma}(j-1)}{C_\gamma^{\frac{j(d-1)}{d}}} \leq c \sum_{j=0}^{\infty} (2^{2g-d+1})^j \overset{!}{<} \infty.$$

In this case the surface to volume ratio of generalised Sierpiński space-filling curve partitions is bounded for $g < (d-1)/2$ and is logarithmically bounded for $g = (d-1)/2$. The case $g = d-1$ has been discussed in Example 2 already. Further results are assembled in Table 4.1. Entries marked by an asterisk * can be computed directly for this example. □

A systematic way to construct adaptively refined meshes is by best n-term approximation. The approximation error is minimised under the restriction that only n degrees of freedom may be used. The idea is to choose these degrees in an optimal way. This is a non-linear approximation procedure. Technically, this best approximation as the solution of a discrete optimisation problem is difficult to achieve. However, there are possibilities to find approximations close to the best approximation in the following way. Given a stable Riesz basis, the size of the weighted coefficients related to an expansion in the stable basis is an indicator of the importance of the basis function. Hence, it is

	$d=2$	$d=3$	$d=4$	$d=5$	$d=6$
$g=0$	$\leq \sqrt{v}$	$\leq v^{2/3}$	$\leq v^{3/4}$	$\leq v^{4/5}$	$\leq v^{5/6}$
$g=1$	$\geq v/\log v$	$\leq v^{2/3}\log v$	$\leq v^{3/4}$	$\leq v^{4/5}$	$\leq v^{5/6}$
$g=2$		$= v$	$= v^*$	$\leq v^{4/5}\log v$	$\leq v^{5/6}$
$g=3$			$= v$	$= v^*$	$= v^*$
$g=4$				$= v$	$= v^*$
$g=5$					$= v$

Table 4.1. *Locality of space-filling curve partitions for adaptive refinement areas of dimension g according to Example 3 with $C_\gamma = 2$. Asymptotic bounds for the surface s up to constants.*

sufficient to choose the n largest weighted coefficients for an approximation, which is up to the stability constants the degrees-of-freedom to error relation of the best n-term approximation, see DeVore [101] and DeVore and Popov [102]. We define the best approximation error as

$$\sigma_{n,t}(g) = \inf_{|\Lambda|\leq n,\ d_\lambda \in \mathbb{R}} \left\| g - \sum_{\lambda \in \Lambda} d_\lambda \psi_\lambda \right\|_{H^t(\Omega)}$$

for a set of basis functions ψ_λ and weights d_λ. We assume that the right hand side of a Poisson problem is in Besov $f \in B^{\alpha-1}_{L_2(\Omega)}$, $\alpha \geq 1$ based on L_2 spaces and smoothness parameter α, see also [101]. Further we assume that we have a stable wavelet basis with at least α vanishing moments. Then we obtain for the procedure choosing the n largest weighted coefficients d_λ

$$\sum_{n=1}^{\infty} (n^{s/d}\sigma_{n,1}(u))^{\frac{s-1}{d}+\frac{1}{2}} < \infty$$

for all $0 < s < \min(\frac{d}{2(d-1)}, \frac{\alpha+1}{3})$, see Dahlke, Dahmen and DeVore [95]. This is the optimal degrees-of-freedom to error rate $n^{-1/d}$ of uniform mesh refinement and H^2 regular problems, as long as $\alpha \geq \frac{d+2}{2d-2}$.

Example 4. *The best n-term approximation gives an explicit characterisation of quasi-optimal meshes. We can study singularities which may occur for the solution of the Poisson equation, see Dauge [99]. Basically, functions of type $r^{-\alpha}$ with $\alpha > 0$ are of interest, which model the behaviour of corner singularities due to the geometry of the domain Ω and due to boundary conditions. The radius r is given in polar coordinates for point singularities or* *e.g. in cyclindrical coordinates for edge singularities. The wavelet coefficients of a representation of the function can be computed explicitly, so that the best n-term meshes are constructed in the following, see also Koster [191]. The convergence rate in L_2 norm is of optimal order $n^{-1/d}$.*

We assume a dyadic mesh refinement (element bisection) and characterise the mesh level by level for a point singularity and $d = 1$. Higher dimensions and other types of refinement can be obtained by direct products of the one-dimensional case and products with uniform refinement. For a given threshold δ, all nodes of a dyadic level $\underline{l}$ and coarser are present. Furthermore, there is a maximum dyadic level $\bar{l}$ and on a dyadic level l in between,

$$n_l = C\delta^{-a}b^l$$

nodes are present, with constants $C, a, b > 0$ and $b < 1$. On each dyadic level only the nodes closest to the singularity are present. Hence the bounds on the levels can be calculated by $n_{\bar{l}} = 1$ and $n_{\underline{l}} = 2^{\underline{l}} + 1$. We parametrise the meshes by a sequence of thresholds, so that we obtain a γ-adaptive family of meshes. For a lower bound of $C_\gamma = 2^d$ we need a sequence of $\underline{l} \in \mathbb{N}$, which results in a geometrically decreasing δ by a factor $(b/2)^{1/a}$. Each mesh τ_i with $\underline{l} = i$ is geometrically graded. The next finer mesh τ_{i+1} can be obtained by a refinement of τ_i. Each element is refined uniformly. Furthermore, a constant number c of elements is created additionally next to the singularity on levels $l > \bar{l}(\tau_i)$. Hence $\bar{\gamma}(i+1) = C_\gamma\bar{\gamma}(i)+c$ and with Lemma 4.19 optimal space-filing curve partitions can be obtained.

For the adaptive mesh refinement toward a g dimensional manifold like for edge singularities ($g = 1$), the upper bound $\bar{\gamma}$ changes. The property

$$\bar{l} = \underline{l}(1 - \frac{\log 2}{\log b})$$

with $b < 1$ holds. Hence more elements are created in a step from τ_i to τ_{i+1} in addition to the uniform refinement. We obtain a g dependent part of the upper bound as

$$\bar{\gamma}(i) = \mathcal{O}(2^{g\bar{l}(i)} + \underline{\gamma}(i)) = \mathcal{O}(2^{ig(1-\frac{\log 2}{\log b})} + 2^{id}).$$

Inserting this into Lemma 4.19, optimal space-filing curve partitions are guaranteed for

$$g < \frac{d - 1}{1 - \frac{\log 2}{\log b}}.$$

In the case of orthogonal wavelets with N vanishing moments and best approximation in $L_2(\Omega)$, the constant b is given by $2^{-(1-\alpha)/(N+\alpha)}$ and the condition reads as $g < (d-1)(1-\alpha)/(N-1)$. This leads to a table analogous to Table 4.1, which depends on α and N.

Note that the Besov regularity for singularities with $g > 0$ of $r^{-\alpha}$ type is much lower than for the point singularity $g = 0$. Hence best n-term approximation does not lead to an $\mathcal{O}(n^{-1/d})$ convergence. This can also explained by the geometric nature of the singularity, where anisotropic functions ψ_λ are more efficient. For similar best n-term approximations results for integral operators we refer to Oswald [233] and Griebel, Oswald and Schiekofer [148]. The approximation results can be improved by anisotropic meshes, which we consider in the next section. $\square$

Now we have several estimates about the partitions of space-filling curves. We can interpret any mesh of a finite number of elements as a quasi-uniform mesh with possibly bad bounds on element volume variation. Furthermore, we can use estimates on β-adaptive refinement where the strength of local mesh refinement enters. However, we know that a large β does not give very useful estimates. In addition γ-adaptive estimates are available where the area of local mesh refinement has to be bounded in a certain way.

Hence numerical experiments can be interpreted in different ways. Qualitatively we get optimal estimates of type $C_{\mathrm{part}} v^{(d-1)/d}$, as long as we do not have an arithmetic progression of elements, see Example 1 and Figures 4.27 and 4.26. The constants C_{part} do depend on the mesh refinement procedure. We see in Figure 4.22, 4.24 and 4.31 that this *constant* is larger for the adaptive case than for the uniform case, which is consistent with theory.

4.3.4 Anisotropic Space-Filling Curves

In Examples 2 and 3 we have seen that the space-filling curve partitions of adaptively refined meshes can give insufficient locality. This will be the case if the mesh has to resolve areas of dimension g too large compared to d, i.e. $g = d - 1$. Although such cases do not occur for elliptic boundary value problems with smooth right hand side, transport or convection-diffusion equations immediately lead to such situations, where boundary layers, shock fronts or jumps of the solution have to be resolved by adaptive mesh refinement. The standard space-filling curves discussed so far lead to partitions with $s = \mathcal{O}(v)$ which leads to too much communication for a parallel implementation. Of course it is possible to distort the space-filling curve anisotropically in order to improve the situation for mildly anisotropic mesh refinement, which might

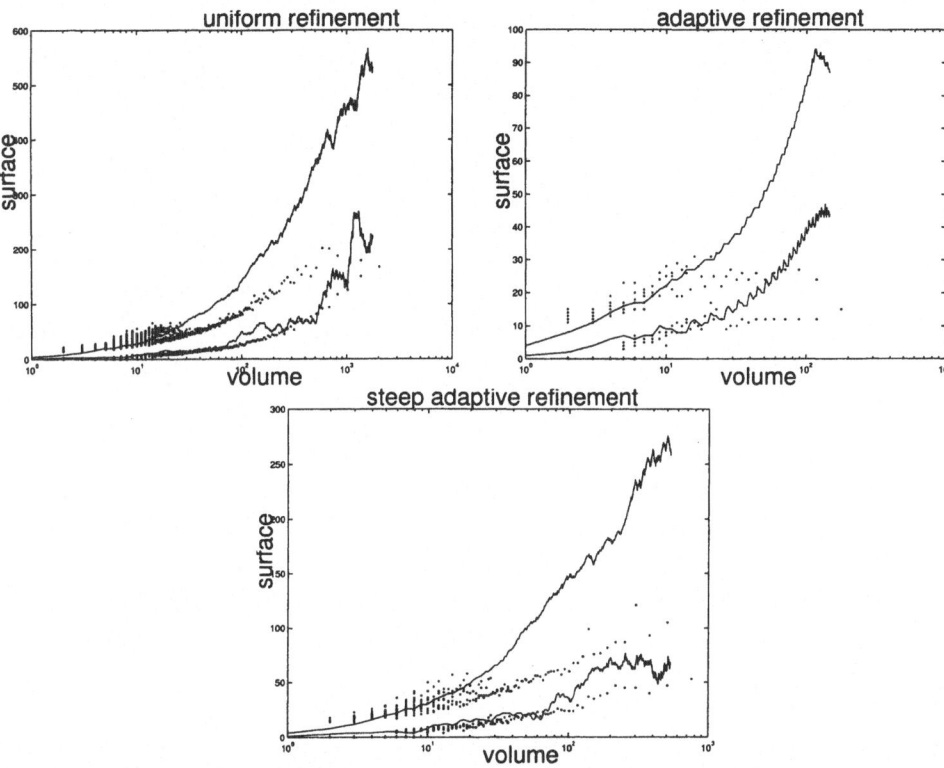

Figure 4.31. *Locality of partitions, smallest and largest surfaces constructed by a Sierpiński curve (solid line) and graph partitioning (Metis, dotted). 3D examples uniform mesh (left), adaptive refined mesh (right) and arithmetically adaptive refined mesh (bottom).*

be sufficient for some convection-diffusion type problems. However, in general a different approach is needed. Analytically, the domain partitioning problem is relatively easy in such cases. Take, for example, adaptive mesh refinement as in Example 2 and in Figures 4.29 (bottom row) and 4.30 (bottom row). A simple partitioning method does help here: cut the domain into vertical slices in the two dimensional case and into tubes in the three dimensional case.

Example 5. *Let us consider adaptive mesh refinement toward an area of interest S of large dimension $g \geq d/2$ where standard space-filling curves give unsatisfactory results of $s = \mathcal{O}(v)$ type. After some local coordinate transformation or a diffeomorphism, we assume adaptive mesh refinement along coordinate axis x_0 to x_{g-1} and standard refinement along axis x_g to x_{d-1}.*

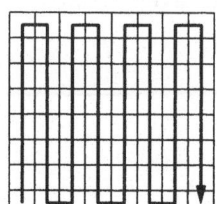 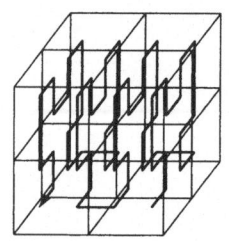 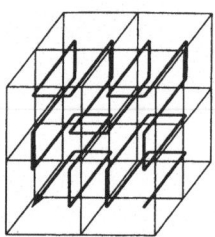

Figure 4.32. *Products of space-filling curves,* anisotropic *curve. Decomposition into* $\mathbb{R} \times \mathbb{R}$ *(left),* $\mathbb{R} \times \mathbb{R}^2$ *(middle) and* $\mathbb{R}^2 \times \mathbb{R}$ *(right) with a two-dimensional Hilbert space-filling curve.*

Now we decompose the domain Ω *into a product of* $\mathbb{R}^g \times \mathbb{R}^{d-g}$. *We define a discrete curve on the product space by space-filling curves on the lower dimensional compact domains in* $\mathbb{R}^g$ *and* $\mathbb{R}^{d-g}$, *where we use the identity as a one-dimensional space-filling curve. The product is constructed such that the curve cycles through all elements of one slice* $(d-g)$-*dimensional before entering the next one.*

The discrete curve, which we will call an *anisotropic* space-filling curve, see Figure 4.32, does not lead to a continuous space-filling curve in the limit case. It can be constructed such that at least the discrete curves are continuous. The standard locality estimates do not hold for the curve. However, the partitions defined by the curve serve our purpose. We can use the locality estimates on the curve orthogonal to the slices to demonstrate the locality of the partitions with respect to the adaptive mesh refinement. In the case of $g = d - 1$ dimensional refinement, i.e. codimension one, the surface of the slices is of the order of adaptive refinement with $g = d - 2$, e.g. sufficient locality as in Example 3 for $d = 2$ with $s \leq \mathcal{O}(\sqrt{v})$ and for $d = 3$ with $s \leq \mathcal{O}(v^{2/3} \log v)$.

General domains can be treated in several ways. We have already discussed the construction of generalised Sierpiński curves on unstructured meshes. These curves can be used for a partition of the domain. However, one can also start with structured meshes in the following ways. The unit square or cube can be mapped to a curved domain by a differentiable mapping function which leads to a variable coefficient problem, several domains can be joined together in an overlapping or non-overlapping way which leads in a natural way to domain decomposition solvers, or the domain can be embedded into the square/ cube with a special type of discretisation. We choose the last version: the domain is enclosed by a cube $\Omega \subset [0, 1]^d$ and all nodes $x \in [0, 1]^d \setminus \Omega$ are considered as boundary nodes. Furthermore, the finite difference stencils are adapted for

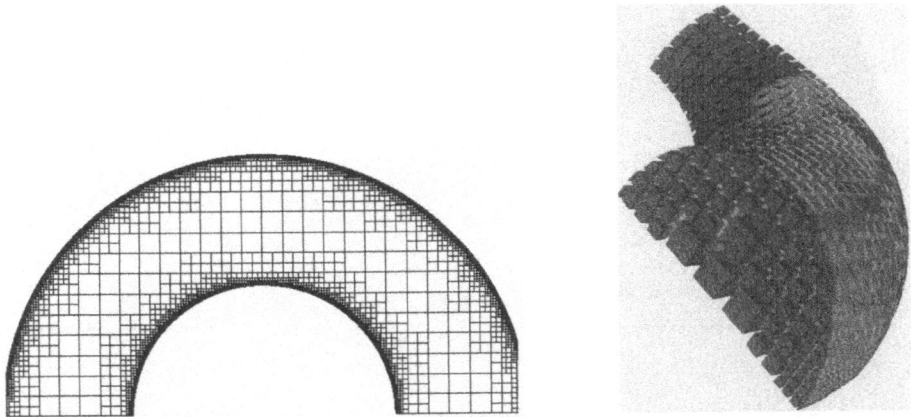

Figure 4.33. *Adaptively refined mesh of a 2D and a 3D torus with a Shortley-Weller type of boundary approximation.*

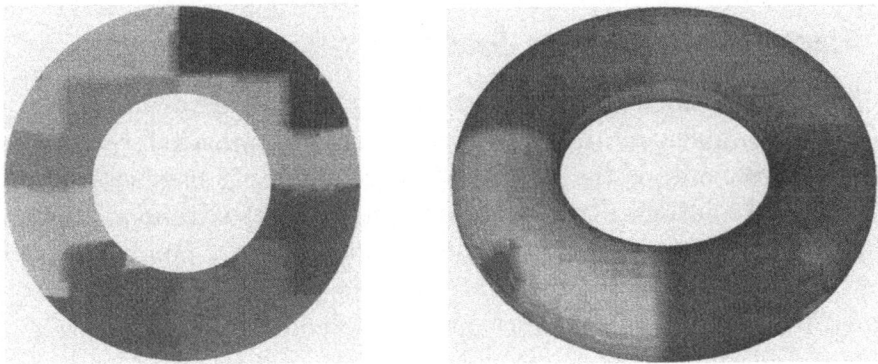

Figure 4.34. *Space-filling curve partitions of meshes of a 2D and a 3D torus for eight processors.*

nodes in the vicinity of the boundary $\partial\Omega$, that is all nodes whose difference stencil crosses the boundary. The step size h is reduced to the actual distance to the boundary, see Shortley and Weller [273] and Hackbusch [168]. Furthermore, adaptive mesh refinement is employed in the vicinity of the boundary to increase the resolution of the boundary, as can be seen in Figure 4.33. We also present the space-filling curve partitions of such meshes in Figure 4.34. Some difficulties related to this approach are the symmetry of the discrete operator and the treatment of certain types of Neumann and Robin boundary conditions in a matrix-free implementation.

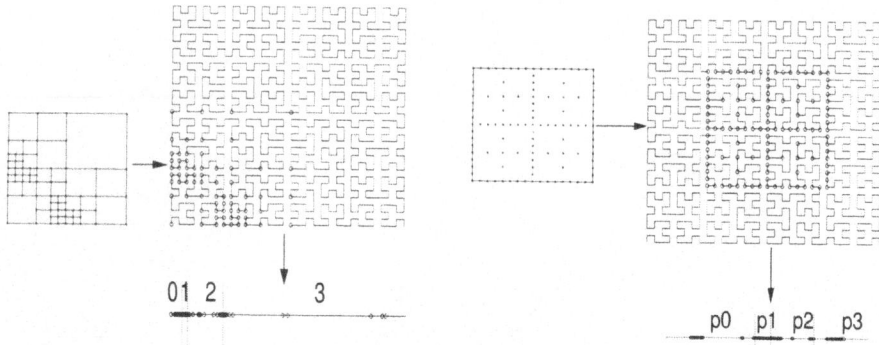

Figure 4.35. *Mapping a 2D adapted mesh to a space-filling curve (left) and a sparse grid to a space-filling curve (right), and mapping points on a space-filling curve to a parallel processors.*

4.4   Partitions of Sparse Grids

Space-filling curve partitions can also be used for the parallelisation of adaptive sparse grid implementations. A very attractive approach for parallelisation of sparse grid schemes is the combination technique. It uses several independently obtained solutions and combines them in a final extrapolation step. The $\mathcal{O}(|\log_2 h|^{d-1})$ grids can be distributed to independently operating solvers on different processors, which gives nearly perfect parallel efficiency rates. However, adaptive grid refinement, non-linearities and other effects cause a tighter coupling of the solvers.

Alternative sparse grid schemes like the Galerkin method or finite differences lead to a more conventional parallelisation so that domain partitioning approaches seem to be attractive. Let us look for space-filling curves for this purpose. We consider a finite difference discretisation which is based on some kind of one-dimensional hierarchical basis transform or (pre-) wavelet transform, its inverse transform and a one-dimensional finite difference stencil. The original concept of nodes and their geometrical neighbourhood has to be extended: the degrees of freedom are still located in the nodes or elements. However, the interaction of nodes becomes non-local. The one-dimensional difference stencils are based on the nearest neighbour nodes along a coordinate axis. The distance of the neighbours is between $h_{\min}$ and $1/2$ for the unit cube $\Omega = [0,1]^d$. The basis transform and inverse transform again uses non-local references. We have to look closer into a generalisation of the term *surface* for a sparse grid algorithm.

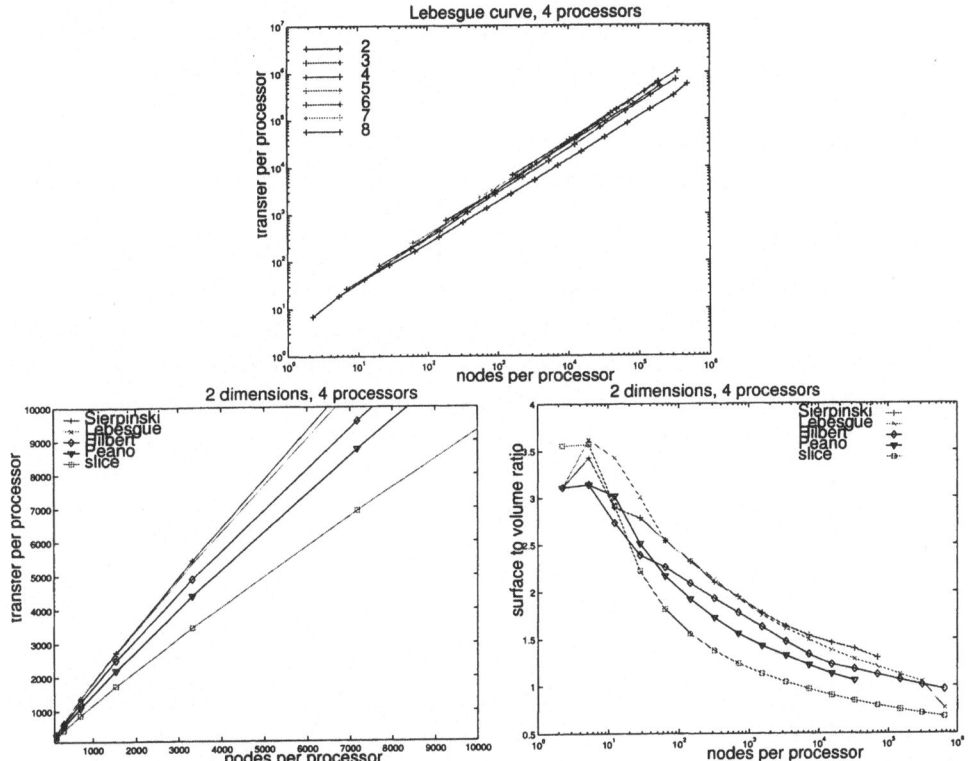

Figure 4.36. *Surface and volume of sparse grid partitions defined by space-filling curves. Performance for different dimensions d with a Lebesgue curve (top). Performance of different two-dimensional curves (left) and the surface-to-volume ratio (right).*

The nodes of a sparse grid can be mapped onto a space-filling curve. The curve can be cut into several pieces which translates back into partitions of the grid, see Figure 4.35. The number of nodes of a regular sparse grid is of the order $n = \mathcal{O}(h^{-1}|\log_2 h|^{d-1})$. The number of nodes of a partition is called *volume*, which is n/p for an even partition onto p subdomains. With *surface* we denote further on the number of nodes to be sent or received in a parallel implementation of a sparse grid algorithm. This is usually more than just the number of geometrical neighbours of the partition, due to non-local references of the algorithms.

A straightforward implementation gives the following numerical results, see Figure 4.36. We obtain a ratio of number of nodes to transfer (surface) to number of nodes owned by a single processor which is slowly decreasing in n roughly like a logarithmic term $1/\log n$. In the previous section standard discretisation schemes usually lead to a surface to volume ratio of $1/\sqrt[d]{n}$. We observe that the logarithmic term is independent of the dimension d and the type of space-filling curve, including the *anisotropic* domain slicing curve. Furthermore, the domain slicing performs slightly better than many of the space-filling curves.

A more detailed statistic shows that the transfers due to finite difference stencils, basis transform and inverse transforms are of the same order. Domain slicing is slightly cheaper than other curves, because the application of difference stencils and basis transform does require data transfer mostly along one coordinate direction, while directions orthogonal need few or no transfers.

Let us analyse the domain slicing partitions. A sparse grid on the unit cube contains $v \approx h^{-1} |\log_2 h|^{d-1}$ nodes. The number of nodes on the surface of the cube is $s \approx 2dh^{-1} |\log_2 h|^{d-2}$ for $d \geq 2$. The surface to volume ratio of the cube is

$$\frac{s}{v} \approx \frac{2d}{|\log_2 h|} \, . \tag{4.9}$$

Furthermore, a sub-cube of side length 2^{-j} has a similar ratio of $2d/|\log_2 h2^j|$ like rectangular shaped domains. We can consider the dependence graph of the finite difference stencils and count the number of cut graph edges. We will get the same ratio, even if the cut positions of the unit cube are not aligned to 2^{-j}. Since the domain slicing performs better than other space-filling curves, we conjecture that the surface-to-volume ratio for all curves is of the order $1/|\log_2 h|$. However, it still might be possible to create partitions differently to improve the surface-to-volume ratio. In order to check this, in a numerical experiment we apply some graph partitioning heuristics to the problem. Since the complete sparse grid discretisation algorithm gives a quite complicated graph, which furthermore depends on the exact parallelisation of the basis transforms, we restrict ourselves to the finite difference stencils. We use the sum of the graphs of all one-dimensional stencils and obtain the results in Figure 4.37.

We see that the graph partitioners are able to improve the best curve partitioning, the one-dimensional domain slicing slightly. However, the general order of the surface-to-volume ratio seems to be the same. A look at the geometric interpretation of the graph partitioner's result indicates that these partitions are also geometric domain partitions. Hence the general estimate Equation (4.9) also applies. According to our parallel computer model Equation (4.4) we obtain a parallel efficiency of

$$\text{efficiency} = 1 \Big/ \Big(1 + \frac{C_2 2d}{C_1} \cdot \log_2 \frac{p}{n}\Big) \, . \tag{4.10}$$

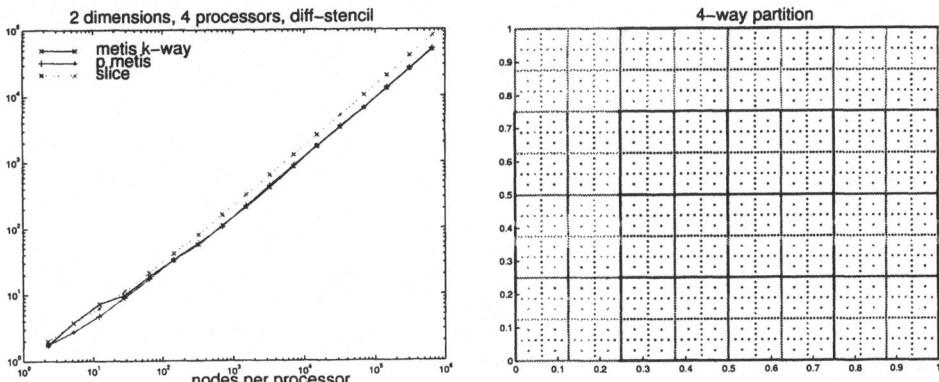

Figure 4.37. *The graph of one-dimensional three-point finite difference stencils on a two-dimensional sparse grid is partitioned into four pieces. Comparison of two graph partitioners with the Lebesgue space-filling curve (left) and the resulting partitions of 4-way Metis (right).*

This implies that sparse grid algorithms are harder to parallelise efficiently than ordinary discretisation schemes based on uniform or adaptive meshes. Parallel efficiency will be high only for very large numbers of nodes. We obtain scalability of wavelet algorithms on sparse grids. However, the parallel efficiency grows far more slowly in the problem size than for standard discretisations, which scale excellently, compare Equations (4.8) and (4.10).

Adaptive mesh refinement is also possible for sparse grids. Guided by some error estimator, additional nodes are inserted into the sparse grid. As long as the structure of the resulting grid is similar to standard sparse grids, we expect similar to worse surface-to-volume ratios. However, if the grid degrades locally to a standard or β- or γ-adaptive mesh, the ratio will improve toward the standard mesh estimates.

In this chapter on space-filling curves we have discussed the partitioning problem. We have introduced classical space-filling curves such as the Hilbert curve, which fills the unit square and unit cube, and Sierpiński-type curves, which we constructed to fill arbitrary triangular and tetrahedra meshes. We have proposed a mesh partitioning scheme based on space-filling curves, which assigns nodes or elements on the image of an interval by the curve to a processor. The remaining chapter dealt with the analysis of the partitioning method. The goal was to derive estimates on the cut size of the graph or the surface of the partition polyhedron in relation to the partition size or volume of the polyhedron. Estimates of this type for uniformly refined meshes were based on the Hölder continuity of the space-filling curve, which guarantees both the connectedness of the partition (neglecting boundary effects) and a relation of the diameter of the partition in relation to the length of the interval. Similar results had been derived earlier by Griebel and Zumbusch [154] and Zumbusch [324]. Furthermore, partitions of this quality are known for uniformly refined meshes and can be constructed easily by geometric means.

However, the main contribution of this chapter is the analysis of space-filling curve partitions for adaptively refined meshes. For this purpose, we have defined families of adaptively refined meshes and the properties of β- and γ-adaptivity. In the first case, a single element in one mesh is refined into a certain number of elements on the next finer mesh. There exist upper and lower bounds of these numbers, which are combined in the single number β. This criterion allows us to generalise the partition estimates of uniformly refined meshes, but does give slightly weaker results, which depend on β. Especially for higher dimensional problems, adaptive mesh refinement cannot be very steep in order to give good partition estimates.

A different mesh refinement criterion describes limits for mesh refinement of several cycles: a single element is refined into a number of elements after i steps and the γ-adaptivity criterion gives upper and lower limits dependent on i. In this way sharp and quasi-optimal estimates on the partitions like for the uniform mesh case are possible. It is shown that standard adaptive mesh refinement procedures indeed comply with this criterion, as does best n-term approximation, as long as the dimension of the singularity manifold is not too large. This means that adaptive mesh refinement toward a point singularity leads to sharp estimates on the quality of the partitions for standard adaptive mesh refinement and for best n-term approximation. Mesh refinement toward an edge singularity in two dimensions may give large cut sizes, which is also shown by a counter example. The same singularity in three dimensions is only a logarithmic term away from the optimum and in dimensions it is optimal, both for standard adaptive refinement and for best n-term approximation in certain parameter regimes. In the case of high dimensional manifolds, anisotropic constructions with space-filling curves are discussed which again lead to optimal cut sizes, if they are aligned properly to the manifold of refinement.

Compared to the analysis of competing graph partitioning methods, the current results seem to be remarkable, since good estimates for balanced graph partitions seem unattainable, in contrast to some known results for unbalanced bisection schemes. Furthermore, space-filling partitioning has a lower sequential and parallel computational complexity than any other graph partitioner with provable quality results. Experimental data in this chapter support the thesis that space-filling curves are competitive from a practical point of view.

Note that the space-filling curve partitioning algorithm is deterministic and incremental, i.e. independent of the results of previous steps but with related results, which lets the program get rid of much of the necessary bookkeeping about neighbourhood relations and data on neighbour processors. Although this is a pragmatic argument, the simplification of a parallel code may be substantial.

The chapter also covers the case of sparse grid partitioning, where simple estimates and numerical experiments show that the best attainable cut sizes are much larger than of standard, local discretisations with compact support stencils or compact support shape functions. Nevertheless, the space-filling curve partitioning can also be applied here. It seems to be even closer to the optimum than for standard discretisation.

Chapter 5

Adaptive Parallel Multilevel Methods

In the previous chapters we have discussed multigrid solvers, adaptive mesh refinement and mesh partitioning by space-filling curves. Now we are in a position to combine these methods. We will have a look at multigrid on a sequence of adaptively refined meshes, parallel multigrid methods, adaptive mesh methods on parallel computers and finally the combination of all three by an adaptive parallel multigrid method.

5.1 Multigrid on Adaptively Refined Meshes

Multigrid methods on adaptively refined meshes are sort of a restriction of the standard multigrid method on uniformly refined meshes. The main difficulty is to maintain the linear complexity of the algorithm. Compared to a sequence of uniformly refined meshes with the same number of levels, the adaptively refined mesh sequence contains fewer degrees of freedom. The remaining question is how to treat the *missing* degrees of freedom. Several suggestions exist, such as patch-wise adaptive refinement by Brandt [64], a sort of local multigrid with smoothers restricted to a neighbourhood of nodes by Rivara [255], hierarchical basis methods by Bank, Dupont and Yserentant [20] and additive multigrid methods by Bramble, Pasciak and Xu [63].

Parallel multigrid methods have been proposed by many authors and in different ways. Basically the sequential scheme is used together with a mesh partitioning scheme. The smoothers and grid transfer operators are restricted to the mesh partition by Brandt [66]. Parallel smoothers like red-black Gauss-Seidel or Jacobi-iterations perform very well in this context. With the introduction of additive multigrid methods by Yserentant [314] and Bramble, Pasciak and Xu [63], a second strategy of parallelisation occurred. Since no smoother is needed in the algorithm, a partition of the coarsest mesh instead of the finest mesh leads to inter-grid transfer operators completely without communication. Hence the implementation is no longer level by level, see for shared memory Leinen [197] and for distributed memory implementations Bey [39] and Zumbusch [321].

The treatment of adaptive meshes on a parallel computer poses many technical difficulties which depend on the structure of the algorithms on the meshes. This holds both for numerical algorithms, which require detailed knowledge of neighbourhood relations and neighbour data, and for mesh manipulation algorithms themselves. Typically the type of adaptivity is restricted in order to simplify the parallelisation as for patch-wise adaptive meshes, where few complete and rectangular shaped blocks of meshes are mapped to processors, see the sequential codes by Berger and Oliger [35] and Berger and Colella [34], and the parallel versions by Bastian, Ferziger, Horton and Volkert [28], Berger and Rigoutsos [36], and Lemke and Quinlan [199]. Unstructured meshes and adaptive refinement on a parallel computer were handled by a library by Williams [307] and a code by Walshaw and Berzins [301], both in the context of time-stepping methods and mesh refinement and de-refinement due to changes in the solution.

The combination of all three components where technical difficulties accumulate, namely element-wise adaptive mesh refinement, multigrid or multilevel solver and a parallelisation thereof, was done later and mostly independent of each other by De Keyser and Roose [100], Bastian [26], Stals [285], and Mitchell [214]. The strategy proposed here extends our concepts of key based addressing over processor domains and hash storage as a parallel hashing. As the major contribution we propose space-filling curves for dynamic load balancing and an implicit computable mapping of data to processors. Together with the estimates on space-filling curve partitions, we are able to prove that this parallelisation does perform well asymptotically. Numerical results will follow in the next chapter.

Originally multigrid and multilevel methods have been defined on uniformly refined meshes. A mesh on a level is assumed as uniform or quasi-uniform and the sequence of meshes is assumed as nested. Hence each element on level l on a mesh is subdivided in the same way into a fixed number of smaller elements on level $l + 1$. There are several ways to generalise this setting to non-quasi-uniform meshes and to adaptively refined meshes, see e.g. [26, 192, 197, 203, 213, 255, 258, 285].

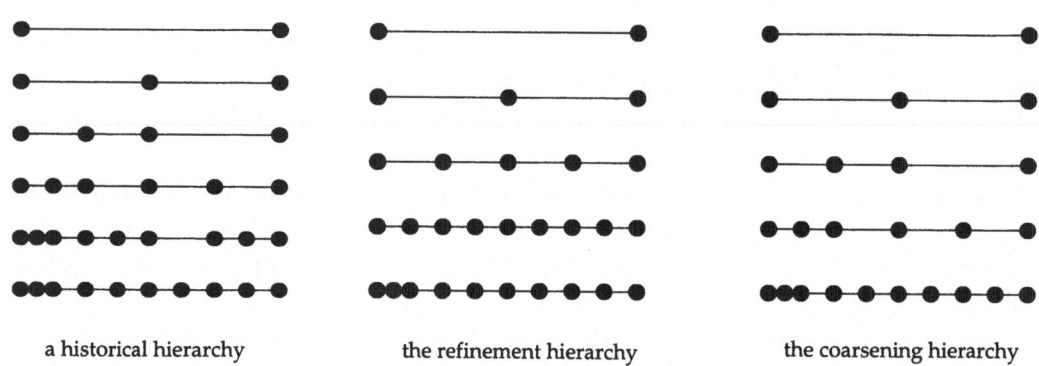

a historical hierarchy   the refinement hierarchy   the coarsening hierarchy

Figure 5.1. *Different mesh hierarchies for a single adaptively refined mesh.*

5.1.1 Mesh Hierarchies

Let us assume that a mesh is created by a successive adaptive mesh refinement, where each mesh refinement step selects a number of elements which are subdivided in a certain manner. The remaining elements remain unchanged. The result of the adaptive refinement procedure is a sequence of nested meshes. We call this the *historical* mesh hierarchy. Alternative mesh hierarchies for the same initial coarse mesh and the same finest mesh can be obtained as follows, see also Figure 5.1: the *refinement* hierarchy starts with the initial coarse mesh. Each element of a mesh is subdivided, if it is not contained in the final mesh and is small enough. Hence, the first mesh levels look very much like uniform refinement, whereas in later levels only a smaller number of elements is subdivided. The mesh hierarchy can also be defined in the opposite way, from fine to coarse. The *coarsening* hierarchy starts with the finest mesh. From one level to the next those elements which are too small for the coarsest mesh are combined to a coarser element in a reverse subdivision process. Now we can reverse the order of the mesh levels from coarse to fine for the coarsening hierarchy, so that all three refinement histories are comparable.

A sequence of uniformly refined meshes can be interpreted as an adaptively refined mesh sequence, where all hierarchies, namely historical, refinement and coarsening hierarchy coincide. For other mesh hierarchies, the three alternatives differ: the number of meshes of the refinement and the coarsening mesh hierarchy is the same and is determined by the number of subdivision steps of the most refined element. In setting of isotropic refinement of elements into 2^d sub-elements, this is determined by the local mesh sizes of the smallest elements of the fine mesh $h_{\min}$ and the element size of the parent element on the coarsest mesh h with $l = 1 + \log_2 h/h_{\min}$. The historical mesh hierarchy has at least the same number of meshes, but may have more mesh levels in cases where the error estimation and mesh refinement process has missed the area of most intensive mesh refinement in a refinement step. The total number of elements or nodes $n = \sum_{i=1}^{l} n_i$ of the mesh hierarchies also differs. The coarsening hierarchy has the lowest number of total degrees of freedom n, because it reduces the amount beginning from the finest level most aggressively. The total number n of the historically mesh depends on the number of meshes, but for a minimum number of meshes the total number of nodes n is at most the total number of the refinement history.

5.1.2 Multigrid

Now that we have a nested mesh hierarchy for an adaptive mesh refinement process, we can apply a standard multigrid method to it. What happens is that we obtain a standard multigrid in areas of locally regular refinement. In areas where elements remain the same from one level to the next level, the prolongation and restriction operators or the associated interpolation operator are locally the identity operator. A smoother in a multiplicative multigrid will be applied to a node several times. The convergence rate of the multiplicative multigrid will not deteriorate due to adaptive mesh refinement. However, the computational complexity of the multigrid method may degrade. The number of degrees of freedom on the finest level n_l may increase less in the number of levels l than the total number of degrees of freedom on all meshes $n = \sum n_i$. In the case of uniform meshes, the sum is just a geometric sum such that

$$n_l \frac{2^d}{2^d - 1} \leq n = \sum_{i=1}^{l} n_i.$$

Hence the complexity of a multigrid cycle $\mathcal{O}(n_l)$ is linear in the degrees of freedom on the finest mesh. However, for adaptive mesh refinement this estimate need not hold, even for a larger multiple of n_l. Consider the case of arithmetic mesh refinement, where only a fixed number of elements is refined from one level to the next. The estimate changes to

$$n_l^2 C \approx n = \sum_{i=1}^{l} n_i.$$

The complexity of a standard multigrid applied to such an extreme case of adaptive mesh refinement is no longer linear, but quadratic in the number of degrees of freedom of the finest mesh, i.e. $\mathcal{O}(n_l^2)$.

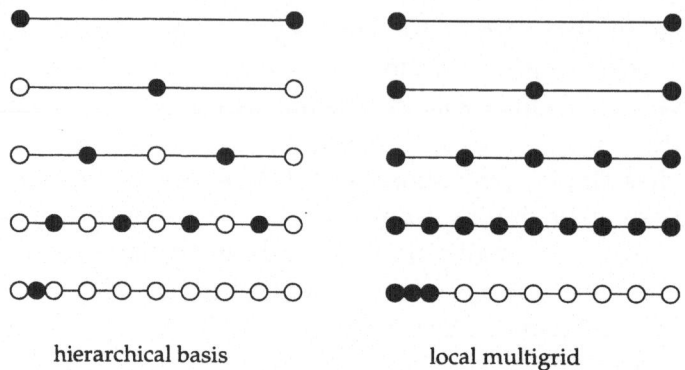

hierarchical basis   local multigrid

Figure 5.2. *Different multilevel methods on adaptively refined meshes. Nodes which are visited or relaxed are marked.*

5.1.3   Hierarchical Basis

In order to circumvent this complexity dilemma, several multigrid versions exist for adaptively refined meshes. One idea is the use of the hierarchical basis, both as a preconditioner by diagonal scaling by Yserentant [314] and Leinen [197] and within a multigrid method as a hierarchical basis multigrid by Bank, Dupont and Yserentant [20] and Bank [18]. The complexity of the algorithm is improved compared to the standard multigrid, because each node is visited only once, namely at the coarsest mesh possible, see Figure 5.2. The amount of work associated with the hierarchical basis method is proportional to the number of degrees of freedom n_l, which is a linear complexity, both for uniform and adaptively refined meshes. The drawback of the hierarchical basis method is the convergence rate or condition number dependence on the finest mesh size or equivalently on the number of levels l. While this dependence is weak in two dimensions with an additional factor l, it is stronger in higher dimensions.

5.1.4   Additive Multigrid

Better convergence rates than the hierarchical basis preconditioner can be obtained by *additive multigrid* methods, namely the BPX-preconditioner by Bramble, Pasciak and Xu [63] and its variants like the multilevel diagonal scaling (MDS) by Zhang [318]. Although the estimate on n holds as for standard multigrid methods, it is possible to implement the methods in linear complexity $\mathcal{O}(n_l)$ for all kinds of adaptively refined meshes, see Bornemann [48]. First of all, the smoothing or diagonal scaling of the additive methods is independent of the order and number of steps. Hence it does not make sense to apply the smoother to all nodes on all levels, but it is sufficient to apply it once per node. Now we can eliminate these parts of the multigrid interpolation procedures which represent the identity operator. The number of operations for a single interpolation on any type of an adaptive mesh will be proportional to the degrees of freedom on the finest mesh n_l, if we consider that each node receives its data from a fixed number of parent nodes one level coarser. The same is true for the complexity of the adjoint restriction operator and hence for the complete additive multigrid cycle.

Note that this complexity cannot be achieved with an arbitrary type of data structure. Data structures which represent the hierarchy of meshes as a collection of meshes suffer a similar complexity problem as they occupy an amount of memory proportional to n instead of n_l. Hence an algorithm which just touches all nodes in the data structure would have a sub-optimal complexity. However, data structures which encode the mesh hierarchy relations within the finest mesh by trees or address-key arithmetic are suitable for such adaptively refined meshes with arithmetic progression.

5.1.5 Local Multigrid

Multiplicative multigrid methods can also be adjusted for an optimal overall complexity as *local multigrid* by a similar procedure as additive multigrid methods, see Rivara [255], Bramble, Pasciak, Wang and Xu [61], and Bastian [26] . Again, a node is relaxed by a smoother originally just once, on the coarsest mesh possible. However, this would directly lead to a hierarchical basis method. In order to maintain the superior convergence rates, the geometric neighbour nodes on that level also need to be relaxed, which is only a constant amount of work, see Figure 5.2. While the naive application of a standard multigrid method would apply a smoother to a node on all levels, the local multigrid reduces this amount. Nevertheless, a node may be relaxed several times, depending on its neighbour nodes on finer levels. Nodes within an area of locally regular refinement are relaxed on all levels, just as in a standard multigrid method. However, in areas where the mesh does not change from level to level, the nodes are not relaxed any longer. The interpolation and adjoint restriction operator can be implemented with linear complexity if only non-trivial operations on nodes required are performed, similar to the additive multigrid methods. Hence the complexity of a local multigrid cycle is optimal $\mathcal{O}(n_l)$, while its convergence rate remains bounded independent of the mesh size as for additive multigrid methods. Again, a suitable data structure for the mesh hierarchy is needed for the optimal complexity of the algorithm.

Note that some adaptive mesh refinement procedures do not lead to nested meshes and nested spaces. This may be caused by the treatment of hanging nodes, if they are not eliminated by interpolation or equivalently by the finite element shape functions. Furthermore, isotropic element refinement by procedures like red-green refinement by Bank, Sherman and Weiser [21] and Bank [18] may lead to difficulties, when an auxiliary green closure is removed in favour of an isotropic red refinement. The finite element spaces of the history hierarchy is not nested on such elements. Although this difficulty can be circumvented by the refinement hierarchy, which is nested, the small perturbation of the multigrid method due to this effect does not seem to be noticeable in practice, see Bornemann, Erdmann and Kornhuber [50].

5.2   Parallel Multilevel Methods

In order to run a numerical simulation on a parallel computer, the implementation of the algorithms has to be modified to take advantage of the parallelism. Independent operations need to be identified which can be assigned to the processors. Furthermore, on a distributed memory parallel computer, data has to be distributed to the processor elements, so that each processor can perform operations on its own data. For reasons of efficiency, the sequences of operations on each processor should be large and the size of data to be exchanged between processors should be small.

5.2.1   An Optimisation Problem

The parallelisation of an algorithm can be formulated as a general graph partitioning problem: each elementary arithmetic operation like '+' or '*' requires some operands and produces a result. The data flow can be represented as a graph with operations as nodes and data dependencies as edges. The minimum execution time of an algorithm is characterised by the shortest or critical path between input layer and output layer, if enough processors are available. However, large scale numerical simulation will always operate in a range where the number of processors is limited and is far smaller than the number of operations and operands. Hence the amount of operations, i.e. the number of nodes per partition, should be the same. For p partitions and n operations, this is n/p operations per partition. Furthermore, the critical path must not exceed the execution time, which is n/p for large n. Now, the optimisation problem is to find the partitions which minimise the number of cuts in the graph, or the maximum number of edge cuts of all partitions. On a shared memory computer, this optimisation problem can be attacked by a dynamic

Figure 5.3. *Partition of a FEM graph to four parallel processors. Partitions are grey coded and edge separators are drawn dotted.*

scheduler, while for the more general case of distributed memory computers a static partition is needed.

This is a fairly general concept. In the case of iterative solvers on meshes, first of all some numerical operations can be combined to a single graph node. Further, the mesh or the sparsity pattern of the stiffness matrix can give a useful hint to the graph or its adjacency matrix. Hence graph partitioning methods are applied to the mesh instead, which leads to a domain decomposition and a partition of data, see e.g. Figure 5.3.

A parallel algorithm on a distributed memory computer consists of sequential parts, where a processor executes ordinary operations on data in its own memory, and communication parts, where at least two processors exchange data, also called communication. In the large scale structured numerical simulations data parallelism is used. This means that all processors execute the same code. Further, all processors either perform sequential operations or communicate with each other. Both the size of data and the number of communication steps has to be small compared to the sequential operations.

Iterative equation solvers, matrix multiplication or the application of a differential operator to a data vector can all be written in a special form: each degree of freedom of the data vector, called a node, is mapped to a single processor. The processor owns a set of nodes exclusively and is responsible for updating data related to the node. The sequential implementation of an algorithm is restricted, so that each processor executes only a set of operations which are necessary to update its own nodes. This is also called *owner computes* paradigm. Hence each node is updated by exactly one processor. In order to perform the necessary operations, some additional operands are needed on the processor. Some of the nodes may have been assigned to other processors. The idea is to introduce additional copies of nodes, so that a processor has access to all necessary operands, either as original or as a copy, also called ghost node. However, for a correct result the content of the ghost node has to coincide with their original nodes at the beginning of the algorithm. This can be achieved by an initial communication step, where each processor requests data of its ghost nodes. In a send-driven communication model, the information, which processor has to send which data for this ghost node update, has to be pre-computed. Now the restricted sequential algorithm can be executed in parallel.

5.2.2 Iterative Solvers

Some standard algorithms parallelise very well: the multiplication by a sparse matrix can be implemented by a single communication step followed by the restricted sequential algorithm. This is a row-wise partition of the matrix onto the processors, see Figure 5.4. The communication pattern for a matrix obtained by discretisation on a mesh is determined by the mesh. The discrete differential operator has a finite support, that is the length of a finite difference stencil or the size of the finite element. For each node in a processor partition, its support is also needed. The partition is geometrically connected or consists of some large chunks of nodes so that most of the supports of the nodes are contained in the partition already. Only a small layer around the boundary of the partition is additionally needed. These nodes have to be allocated as ghost nodes. Hence the discrete support and the size of boundary of the partition determine the number of ghost nodes, which is a direct measure for the size of data to be communicated. The number of neighbour processors within a communication step can be characterised geometrically again: it is the number of neighbour partitions with respect to the sparsity pattern of the discrete differential operator, which is equivalent to the mesh in many cases. Hence the matrix multiplication needs one local communication step to fill the ghost nodes and a sequential computation phase, which can be run without further overhead due to parallelisation.

In a similar way as in matrix multiplication, a single step of a Jacobi iteration can be implemented on a parallel computer, see Figure 5.5. The computation of the residual has exactly the same characteristic as the matrix multiplication. The additional diagonal scaling and vector operations do not

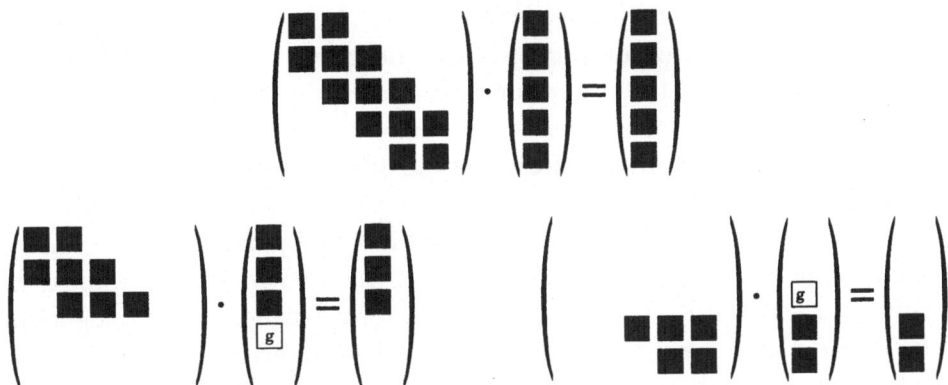

Figure 5.4. *A schematic sequential matrix multiplication for a one-dimensional three-point stencil or linear finite elements (top). A parallel matrix multiplication with row distribution of the matrix on two processors (bottom left and right). The values of the ghostnodes marked 'g' have to be filled (copied) in advance by a communication operation.*

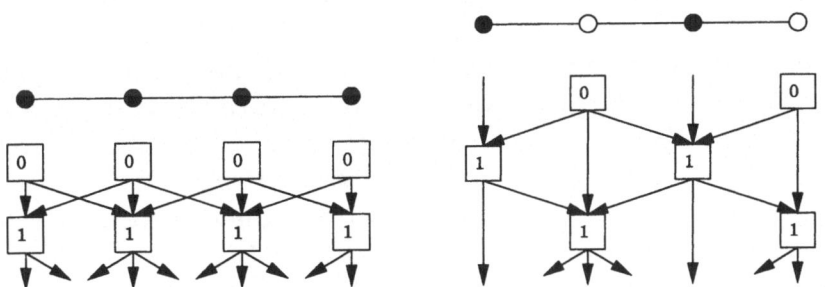

Figure 5.5. *Data flow pattern of a single Jacobi iteration step (left) and a Gauss-Seidel iteration step with red-black node ordering (right).*

require ghost nodes and do not cause further communication. However, the use of several Jacobi steps leads to a nearest neighbour communication in each iteration, followed by a sequential computation step.

Theoretically it is possible to reduce the number of communication steps for multiple iterations. This can be achieved by a number of layers of ghost nodes around the processor partition, so that multiple restricted sequential Jacobi steps still deliver correct results. However, in this case more computation has to be performed on each processor, since for each additional iteration an additional layer of ghost nodes needs to be filled. In order to ensure correctness, these layers also require computation on the ghost nodes, so that the last iteration is restricted to the processor's own nodes and is equivalent to a single step Jacobi iteration.

Iterative schemes of multiplicative type like the Gauss-Seidel and SOR iterations are harder to parallelise. Due to a prescribed order of nodes, a dependence of the nodes exists which may prohibit independent operations completely, see the symmetric Gauss-Seidel iteration with lexicographical node ordering in Figure 5.6. However, other node orderings lead to algorithms, which can be parallelised easily. One way to characterise these dependencies is a colouring of nodes. This is again a graph problem, however of lower complexity than the graph partitioning problem. Each node gets a colour so that two adjacent nodes always have different colours. The graph is determined by the dependencies of the algorithm. In the case of the Gauss-Seidel iteration it is simply the sparsity pattern of the stiffness matrix. We look for a node colouring with a minimum number of total colours which is also called the *chromatic* number of the dependency graph. For example, a five-point stencil on a structured, two-dimensional mesh allows for a two-colour colouring, namely the chequerboard or red-black colouring, see Figure 5.5. The quadratic nine-point stencil or (bi-)linear finite element discretisation leads to a four-colour colouring. The idea is that the nodes of one colour are independent from each other. This means that a parallelisation of the iteration can be based on a local neighbour communication step and a sequential computation step per colour. A j-colour colouring leads to j communication steps. Hence, the Gauss-Seidel iteration requires at least two times the number of communication steps than the Jacobi iteration. The size of data to be exchanged may be larger for the Gauss-Seidel, but can also be the same as for the Jacobi iteration, which depends on the discrete operator. Again, multiple iterations may allow for an optimised communication pattern, see Figure 5.6. Note that colouring schemes can also be used to parallelise algorithms like multilevel methods dynamically on shared memory computers, see also Leinen [197].

Within a multigrid method, a fast smoother may be required. Hence a combination of the good communication pattern of the Jacobi iteration with the superior convergence rate of the Gauss-Seidel or SOR method is sought. The idea is to restrict the communication to a single nearest neighbour step, but to use an order dependent scheme within each processor partition. This can be interpreted as an outer Jacobi iteration with a Gauss-Seidel iteration inside. However, other variations with more expensive solvers on the processor partition are possible. Even an exact solution by a direct method can be used, which is equivalent to an overlapping Schwarz method with minimum overlap size. This small overlap leads to a poor convergence rate for small mesh sizes h.

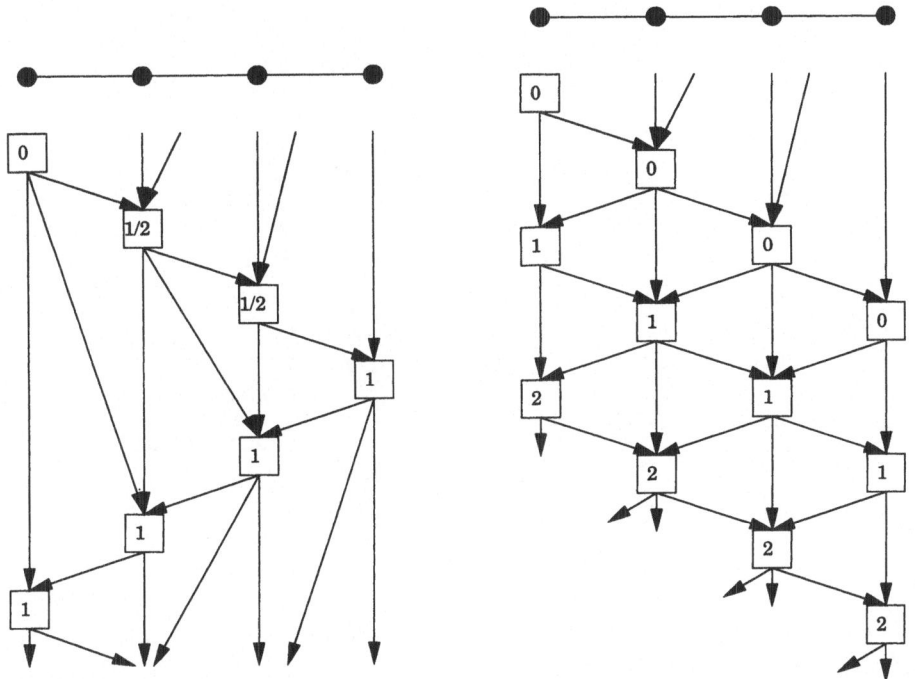

Figure 5.6. *Data flow pattern of Gauss-Seidel iterations with lexicographic one-dimensional node ordering: a symmetric Gauss-Seidel iteration step (left) and a sequence of Gauss-Seidel steps (right).*

5.2.3 Krylov Solvers

Krylov iterative methods can be parallelised in a straightforward manner. The main operation is the application of the operator and, for the preconditioned versions, the application of the preconditioner. Given some parallel versions of the operators, for example as parallel matrix multiplications, most of the vector operations do not require further communication. The scalar products cause some communication, since the local processor sums have to be combined to a single global value. This can be done by a *reduction* operation. Its optimal implementation is hardware dependent, but can be based on a binary tree of processors, both to collect and combine data and to distribute the result, see also Figure 5.7. Hence the scalar products have a parallel complexity of $\mathcal{O}(n/p + \log p)$, which differs by the logarithmic term from other estimates. Within the GMRes iteration [128] further communication is needed for the orthogonalisation. Modifications of the conjugate gradient method exist where both scalar products can be computed at the same time, which helps to save

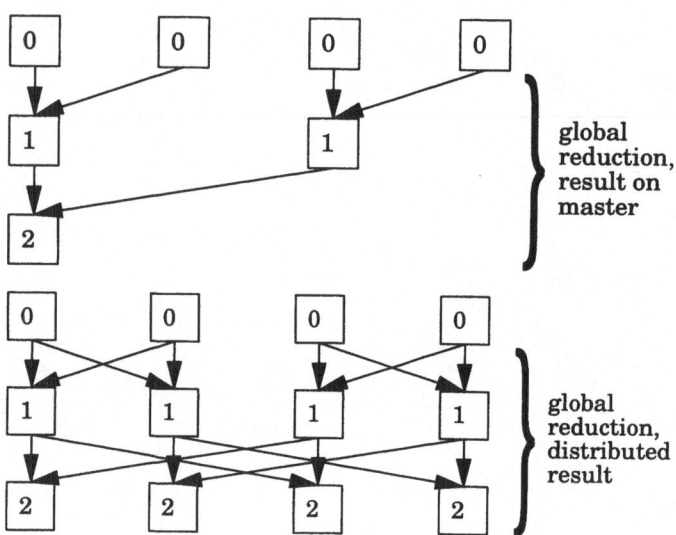

Figure 5.7. *Communication pattern of a global reduction operation, e.g. for the evaluation of a scalar product: a binary tree of depth $\log_2 p$ so that the result is available on the master processor (top) and a butterfly graph as intertwined binary trees so that the result is available on all processors (bottom).*

the communication overhead of one of the reduction operators. The modified method may be less stable than the original version.

Furthermore, within the setting of Krylov solvers it is possible to change the communication pattern of the iterative solver so that a matrix multiplication can be performed without communication, see Bastian [26] and Haase [159]. However, this communication cannot be eliminated, it can only be postponed. This is possible because of the linearity of the vector operations, including the scalar products. The idea is to use a semi-assembled stiffness matrix, where the local stiffness matrices are assembled on the processor partition only, as in the Neumann-Neumann preconditioner for non-overlapping domain decomposition, see section 2.4.3. The final communication step of the global matrix assembly step is omitted. This matrix representation holds some rows and columns several times, so that the global matrix can be computed as the sum of these rows and columns. This *semi-assembled* matrix storage, see Figure 5.9, does not permit a communication free matrix multiplication. However, the multiplication by an ordinary vector leads to a semi-assembled vector, similar to a column-distributed matrix, see Figure 5.8. Usually this

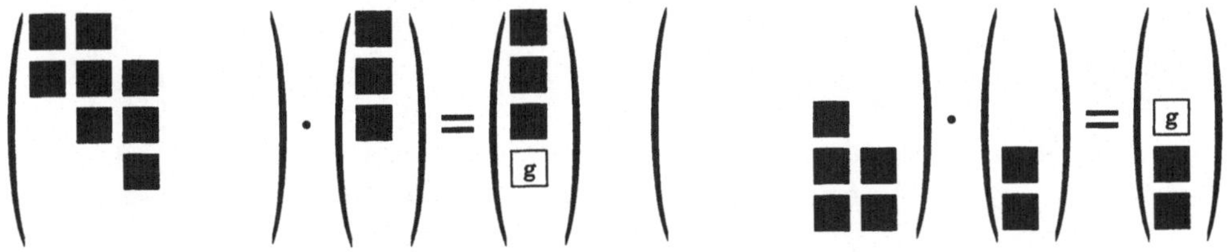

Figure 5.8. *A parallel matrix multiplication with column distribution of the matrix on two processors (left and right). The values of the ghostnodes marked 'g' have to be updated (summed up) afterwards by a communication operation.*

matrix multiplication would be followed by a communication operation. Now this communication is postponed and integrated into the preconditioning step. The result of this operation is that, aside from the global reduction operations for the scalar products, all inter-processor communication can be concentrated within the preconditioner. The total communication depends on the communication of the original preconditioner, which can be zero for a diagonal scaling (single Jacobi step) or large for a multiplicative multigrid. The amount of communication of the iterative solver is comparable to *owner compute* parallelisation strategies, but the amount of memory and amount of sequential operations is slightly larger, depending on the size of the processor partition boundaries. The main advantage of the *semi-assembly* is the elimination of the single communication step in the matrix assembly. However the bookkeeping, which vector in an algorithm is *semi-assembled* and which is stored in the standard way, can be confusing.

5.2.4 Multilevel Methods

For the parallelisation of a multilevel method, several components have to be discussed and various strategies are available. The largest fraction of operations of a multilevel method uses data on the fine mesh level. The finest mesh can be partitioned, so that parallel versions of a smoother like block Jacobi variants, restriction and prolongation operators are applicable. However, this is not sufficient, as part of the computational work has to be performed on the coarser meshes which cannot be neglected. Hence a partition of the coarser mesh levels is also necessary. This partition has to minimise both processor communication for operations on the specific mesh level and, at the same time, communication during the inter-grid transfer operations.

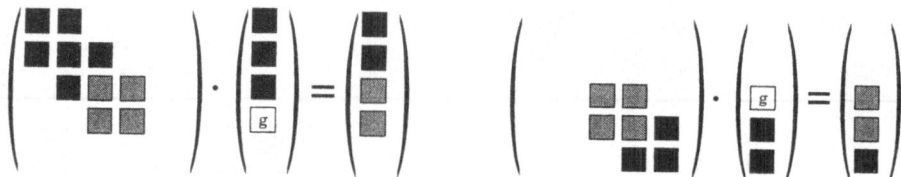

Figure 5.9. *A parallel matrix multiplication with a semi-assembled matrix on two processors (left and right). The values of the ghostnodes marked 'g' have to be filled (copied) in advance by a communication operation. The result vector is also* semi-assembled. *In order to convert it to a consistent vector, an additional communication operation (sum up) is needed.*

At some coarse level it may be suitable to reduce the number of active processors. The coarsest mesh can be handled on a single processor. This causes additional communication because the inter-grid transfers are more expensive. Alternatively, identical coarse mesh computations can be replicated on all processors, so that no processor is idle. However, the communication cost of a concentration of data remains similar. Although the number of operations on coarse meshes is small, the parallelisation requires some care.

Multiplicative algorithms require the execution of the constituents of the algorithms one after the other sequentially. The respective additive versions allow for a parallel execution of the same parts. This is true for the multigrid cycle, where an additive method allows to combine operations and communication of inter-grid transfer operators and scaling, see Figure 5.10. This is also true for smoothers, where additive versions are easier to use than multiplicative versions. A parallel version of multiplicative multigrid is usually based on a partition of all nested meshes. The domain Ω is decomposed into several subdomains Ω_j, which induces partitions of all meshes. Each processor holds a fraction of each mesh in such a way that these fractions of each mesh form a nested sequence. Hence each operation on a specific level is partitioned and mapped to all processors. Furthermore, the communication during mesh transfer operations is small because of nested sequences on a processor. This means that one has to treat global problems, which are partitioned to all processors, on each level. The intra-grid communication has to be small, that is the number of nodes on the boundary of the partition should be small.

A static partition of the domain into strips or squares can be used for uniform meshes and has been used for the first parallel multigrid implementations

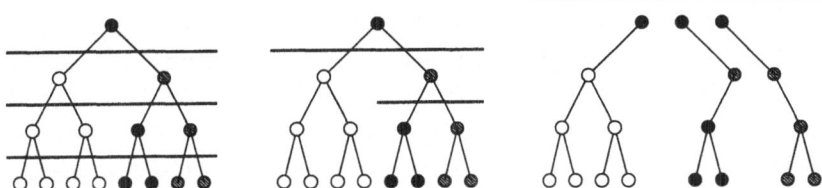

Figure 5.10. *Parallel multigrid methods. A level-wise parallelisation of a multiplicative multigrid method with a global communication step between two levels (left); An additive multigrid method with communication, whenever parent and children nodes are on different processors (middle); an additive multigrid method with a single communication step, but some replicated data (ghost nodes) (right).*

by Grosch [155], Brandt [66], and Solchenbach, Thole and Trottenberg [283], see also the survey by McBryan et al. [208]. Unstructured meshes can be partitioned by some graph partitioning methods. In order to minimise the communication of the inter-grid transfer operators, the partitions of the different mesh levels should be related. A straightforward way to do this is to partition a single mesh and to derive the partitions of the remaining mesh levels from this partition. For large coarse meshes, it may be sufficient to partition the coarse mesh. Each element on a finer mesh level can always be mapped onto the same processor as its parent element. Hence, there is no communication for the restriction and prolongation operation. However, the resulting partition of the finest mesh may be too crude, so that a partition of a finer mesh or even the finest mesh may be preferable. In such cases, coarser mesh partitions have to be derived from a finer mesh partition. An element can be mapped to the same processor as one of its children elements. This mapping is not unique but sufficiently good for the coarse mesh operations. Note that in this case communication occurs for restriction and interpolation on elements which have children elements on different processors.

In contrast to the geometry oriented parallelisation of multiplicative multigrid methods, the additive multigrid version or additive multilevel preconditioners can be parallelised in a more flexible way. The overall workload has to be partitioned, but we do not have to consider individual levels. Here the communication can also take place in a single step for all nodes which are located on the boundary of at least one mesh of the nested sequence. This is efficiently possible if again the coarsest mesh is partitioned and finer meshes inherit the partition. In other cases, the amount of extra work due to a single communication step increases, the finer the partition is. For finer mesh partitions, either more communication steps or some additional operations are necessary, see Figure 5.10. The quality of the multilevel partition has to be traded against communication of the inter-grid transfer operations. The multilevel BPX preconditioner for a uniform mesh has been parallelised by Bey [39] and Zumbusch [321]. These approaches can easily be generalised to block-structured meshes.

There are many suggestions to modify components of the multigrid algorithm for parallel computing.. Multiple coarse meshes can keep all processors busy during operations on coarse meshes, where often some processors are idle. Parallel versions of the smoothers, which take the partition into account, enhance the parallel performance. The usual communication steps during the execution of the smoother on one level can be reduced, if one uses block Jacobi-type smoothers. Some other variants such as a parallel point- and domain-block re-formulation were proposed by Griebel [138, 139] and Griebel and Neunhoeffer [143]. An analogous parallelisation method was developed by Brandt and Diskin [69]. For a more popular introduction to parallel multigrid versions see Douglas [106]. More recent modifications of multigrid algorithms are concerned with the performance on computers with memory hierarchies, such as RISC processors with memory caches. The execution order of inter- and intra-grid transfer operations has to be reordered in order to minimise the required memory bandwidth, see Douglas [107] and Stals and Rüde [286].

5.3   Parallel Adaptive Methods

Parallel algorithms developed for unstructured meshes and data sets can also be used for adaptively refined meshes, see Figure 5.11. However, some of the difficulties of parallel multigrid methods on adaptively refined meshes can also be found for other parallel algorithms on an adaptively changing data set.

The partition of data has to be found at run-time. The adaptive mesh refinement technique creates meshes, one after the other. The adaptively refined meshes are part of the characterisation of the solution and are generally not known in advance. The amount of computational work for the data partitioning should not exceed the work for the numerical algorithm under consideration.

Furthermore, due to the amount of data, the partitioning procedures need to run on the parallel computer. Both memory and network resources do not allow to run a mesh partitioning algorithm on the full data set on a single processor.

The adaptive refinement algorithm modifies the mesh in use. Hence load imbalance occurs and a new data partition is needed. However, this partition

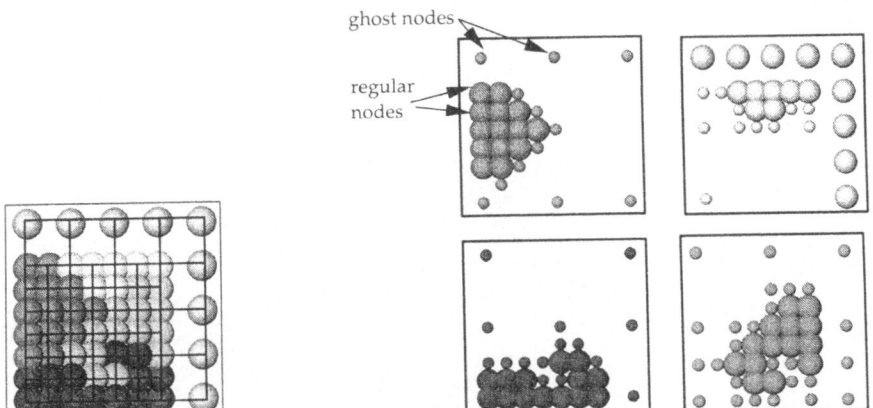

Figure 5.11. *An adaptively refined mesh partitioned into four parts. Regular nodes and ghost nodes for a five-point finite difference stencil.*

has to be related to the previous partition in order to avoid large data movements due to the partitioning or data migration phase. Such heuristics are also called iterative partitioning methods.

5.3.1 Multigrid

The parallelisation of multigrid methods on adaptively refined meshes can be done similarly to parallel multigrid methods on uniformly refined meshes. The basic ingredients for multiplicative methods are again parallel smoothers and parallel inter-grid transfer operators. Additive methods rely on parallel inter-grid transfers for more general partitions of nodes, see Figure 5.12. Special versions of multigrid methods for adaptively refined meshes like local multigrid can be parallelised similarly. However, the main difficulty in this respect is the partition of the computational work. It is no longer sufficient to partition meshes.

- A single mesh partition can be projected to the other mesh levels. However, the load balance on some of the meshes will be poor due to the adaptive mesh refinement.

- Partitions of several meshes can be optimal for operations on the specific level. They can take the amount of work related to local multigrid methods into account. However, the inter-grid transfer operations have to restrict and interpolate data between mesh levels, so that the mesh partitions have to be related.

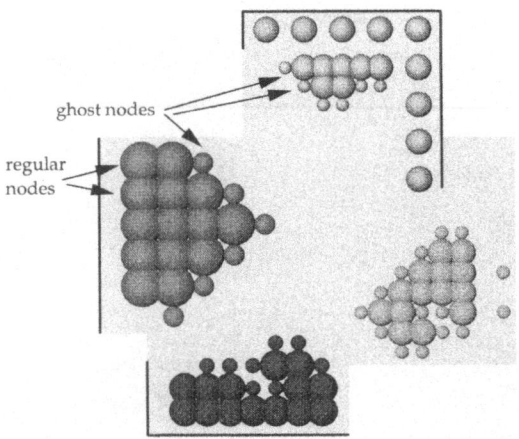

Figure 5.12. *An adaptively refined mesh partitioned into four parts. Regular nodes and ghost nodes for an additive multigrid method.*

- A partition of a previous sequence of adaptively refined meshes need not be optimal for the new sequence of meshes created by an additional single adaptive refinement step. A re-partitioning of the meshes is needed.

5.3.2   Scheduling and Partitioning

The mesh partition problem can be equivalently formulated as a graph partitioning problem. However, the general graph partitioning problem is NP-hard. Even the problem to find asymptotically optimal partitions for unstructured meshes is NP-hard, see Bui and Jones [76]. Several heuristic algorithms have been developed in the area of parallel computing: there are bisection algorithms based on the coordinates of nodes and elements and there are many algorithms based on the graph of the stiffness matrix. For instance, Pothen, Simon and Liou [245] propose a spectral bisection method based on the discrete Laplacian of the graph. The Fiedler vector of the graph is used to find a bisection with a short cut. While an approximation of the eigenvector is sufficient and iterative Lanczos-type methods can be used, the computation can be improved by a multigrid or nested iteration type method, see Barnard and Simon [24]. In the package Chaco by Hendrickson and Leland [174] the graph is cut into several parts by an eigenvector of a larger eigenvalue than the eigenvalue of the Fiedler vector. The package Metis by Karypis and Kumar [183] uses a multilevel version of the Kernighan-Lin algorithm, which improves a graph partition by moving nodes from one partition to another. Its multilevel version does this also on coarser representations of the graph with a heuristic graph coarsening.

In addition to static data partitioning, dynamic or incremental re-partitioning is also needed within an adaptive mesh refinement setting. A large class of dynamic algorithms is based on the data diffusion approach. Data is moved in synchronous steps from a processors to neighbour processors with lower node. This is an iterative procedure, where load imbalance is reduced step by step. Basically, it is a Jacobi-type of iteration on the discrete Laplacian and has been analysed by Cybenko [94]. The convergence rate depends on the network and number of processors. The convergence rate of the data diffusion can be increased by over-relaxation analysed by Ghosh, Muthukrishnan and Schultz [127] and a multigrid-type of iteration by Horton [178]. In addition to the problem of convergence of the data diffusion itself, which always leads to an equidistributed workload, there is the question of how to partition a given fraction of a mesh. Only the number of nodes or elements to be transferred in a single diffusion step is prescribed by the algorithm. Diffusion methods tend to give partitions with large partition boundaries and with partitions consisting of multiple non-connected meshes. Furthermore, the amount of data which is transferred within the data diffusion algorithm is larger than of methods optimised for that purpose. Data diffusion heuristics are employed in Jostle by Walshaw and Berzins [301], in the PDE codes 'UG' by Bastian, Eckstein and Lang [29] and in codes by Stals [285], Berger and Bokhari [33], and Bokhari, Crockett and Nicol [46]. For a survey on mesh partitioning methods we also refer to Pothen [244].

The data partitioning problem can be handled in different ways, which also depends on the details of the methods. Early implementations of adaptive multigrid methods are based on the simple uniform mesh case: the adaptive mesh is composed of the union of uniform rectangular meshes of different area and mesh sizes h. During adaptive refinement, new patches are created and overlaid on the existing union of patches where mesh refinement is needed. This technique of adaptive mesh refinement is popular for transport problems and hyperbolic conservation laws and called 'AMR' (adaptive mesh refinement) in this context, see Berger and Oliger [35] and Berger and Colella [34]. Algorithmic difficulties occur when single nodes, which are flagged for adaptive refinement, have to be combined into a small set of rectangular shaped patches, so that only a few non-flagged nodes are refined additionally. Sequential and parallel heuristics exist for this purpose, see Berger and Rigoutsos [36].

A multigrid method on such a composite mesh can be parallelised in two ways: either each patch is assigned to a single processor, or each patch is partitioned and mapped to all processors in the same way as in the uniform mesh case. This has been implemented on distributed shared memory architectures by Bastian, Ferziger, Horton and Volkert [28]. On distributed memory computers, several packages have been developed so far, such as a multigrid implementation for 'LPARX' by Kohn and Baden [187], 'AMR' by Lemke and Quinlan [199], Quinlan [248], and Lemke [198] based on a parallel array package, and the block-structured Navier-Stokes solver LiSS by Ritzdorf and Stüben [254]. Later AMR packages and computing environments include 'Chombo', 'Samrai', a code by Lötzbeyer and Rüde [204] and others. The distribution of patches onto processors can be written as a knapsack optimisation problem, where even simple heuristics give acceptable results. The strategies can be run both dynamically in a shared memory environment and statically in a distributed environment. Each mesh level or layer of patches has to be distributed independently for multiplicative multigrid methods.

An element-wise adaptive mesh refinement requires more detailed mesh partitioning methods. These can be based on graph partitioning heuristics or on geometric heuristics and can be expensive. However, in the framework of adaptive element-wise refinement, partitions and mappings have to be computed quickly and during run-time. On a shared memory parallel computer, the serial representation of the mesh hierarchy in memory and colouring of the elements along with dynamic scheduling of lists of elements can be used. This has been proposed for a code based on triangles and the additive hierarchical bases preconditioner by Leinen [197] and Birken [42]. Space-filling curves have been used for scheduling by Heber, Gao, and Biswas [171].

On a distributed memory computer, the mesh hierarchy has to be partitioned and maintained, which requires a substantial amount of bookkeeping. De Keyser and Roose [100] propose a multiplicative multigrid method or a finite volume discretisation on a mesh with quadrilaterals and hanging nodes. The elements are partitioned by a hierarchical recursive coordinate bisection. Initial numerical experiments in the package 'UG' by Bastian [26] indicated that the re-partitioning of coarser levels when new elements were created did not pay off. Adaptive conforming meshes consisting of triangles, which are refined by a bisection strategy, were employed in the parallel multigrid codes by Stals [285] and Mitchell [214]. Here, different re-partitioning strategies for refined meshes were used.

The space-filling curve partitioning of chapter 4 can be used in several different ways in this context. First of all, we know how to partition a given single mesh. Furthermore, we have estimates on the quality of the partitions and we know that the surface to volume ratios of the partitions are of quasi-optimal order under some assumptions on the meshes. Hence, a single mesh iterative scheme can be parallelised well.

The partitioner itself is parallel and can be implemented e.g. by a parallel single step bucket sort. Hence we can partition and re-partition a mesh quickly. It is easy to see that small changes of the mesh lead to small changes in the partition due to the properties of space-filling curves. This is e.g. not the case for many graph-based partitioners. Hence the space-filling curve method can be used as an incremental partitioner, which is used every time the mesh is changed. Of course, re-partitioning can also be expensive if the mesh has changed a lot.

As we saw in this chapter, there are several versions of parallel multigrid methods. Standard multiplicative multigrid methods on regular or quasi-uniform meshes have to be parallelised level by level. However, the space-filling curve partition of each level individually and the partition of the finest mesh, which induces partitions of the coarser meshes, do not differ substantially. This is due to the self-similarity of the space-filling curves. The communication at the inter-grid transfer operations is small.

An additive multigrid method on the same sequence of meshes can be parallelised by a partition of the coarsest mesh. However, using space-filling curves this can be interpreted as a space-filling curve partition of each level with certain interval partitions. Alternatively, the additive multigrid method can be parallelised with a larger number of communication steps similar to the multiplicative method.

Parallel multigrid methods on adaptively refined sequences of meshes require a different parallelisation strategy. The additive method can still be parallelised by a partition of the coarsest mesh, as long as the granularity of this partition is fine enough. Again this can be interpreted as a space-filling curve partition of all meshes. A small perturbation of this partition leads to an arbitrary partition of the finest mesh with a different parallel communication pattern. We still have the estimates on space-filling curve partitions on a single level. We can sum up the estimates over all levels and get an estimate for the additive multigrid method.

The multiplicative multigrid of choice on adaptively refined meshes is the *local* multigrid method. The active nodes rather than the meshes on each level have to be partitioned. This can be done by space-filling curves, which give good partitions of each level as long as the set of active nodes is well shaped. However, the question arises on the communication of the inter-grid transfer operators. Some arguments show that this communication cannot be bounded in general. Hence a slightly different strategy may be appropriate. A node clustering with small trees of several levels is used for the partitioning, see Bastian [26]. Alternatively, a space-filling curve partition can be constructed so that each node is mapped to a processor according to a partition of its finest level. We have a different partition on each level which maps only part of the nodes. Hence the parallel inter-grid transfer is cheap. The communication of the inter-grid transfer within each refinement region is also cheap. However, the overall communication cost of the inter-grid operator depends on the refinement pattern. This is one reason, why we use additive methods in the following.

The deterministic behaviour of the partitioner is used in the next section. Sometimes it is extremely useful to be able to compute the owner of some data. Imagine the situation where some neighbour nodes are needed. Rather than asking all processors, the knowledge of the partition of the interval, which comprises to $p - 1$ numbers, and the space-filling curve allow to compute the answer. Thus some type bookkeeping concerning the parallel data mapping within a hash storage scheme can be avoided.

5.3.3   Parallel Hash Storage

The parallelisation of an adaptive code usually is non-trivial and requires a substantial amount of code for the parallelisation only. Hierarchies of refined meshes, where neighbour elements may reside on different processors, have to be managed, see Birken [43]. In particular, dangling pointers have to be avoided. That is, appropriate ghost nodes and elements have to be created and updated when the parallel algorithm performs a communication operation. Furthermore, tree data structures store relations between neighbour elements, parent and children elements, elements and nodes and perhaps more. This happens both in the numerical part, where an equation system is set up and solved, and in the non-numerical part, where meshes are refined and partitioned, see also [26, 29, 100, 181, 214, 271, 285, 307].

In section 3.3.6 we have considered hash storage techniques to simplify the implementation of a sequential adaptive code. Now, we generalise the concept of key addressing and hash tables to the parallel case. The idea is to store the data in a hash table located on the local processor. However, we use global keys, so a ghost copy of the node may also reside in the hash table of a neighbour processor. Furthermore, we base the code on a mapping key-to-processor. We do this by space-filling curves, which leads to space-filling partitions, but other mappings are possible. The position of a node on the space-filling curve, along with the known partition, defines the home processor of a node, see Figure 5.13. Given an arbitrary node, it is easy to determine to which processor the node belongs to. If a node is found that does not belong to the processor, it must be a ghost copy, and the processor can compute where

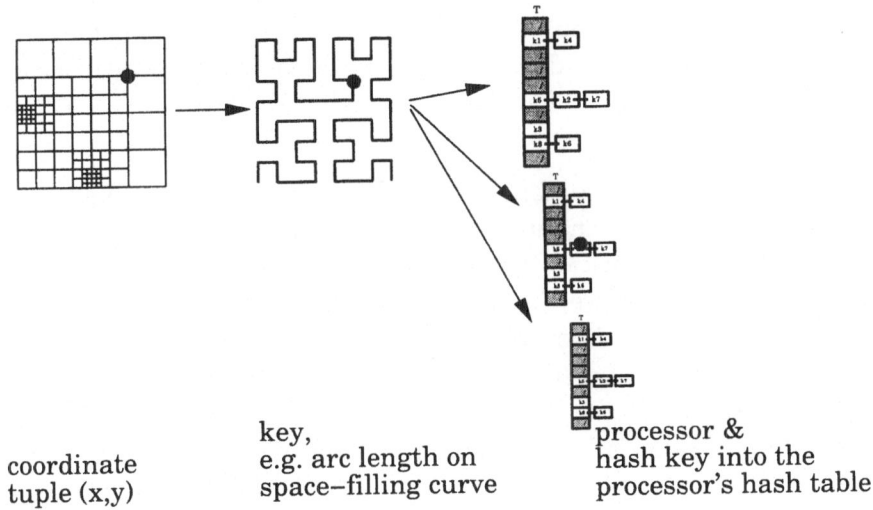

coordinate
tuple (x,y)

key,
e.g. arc length on
space–filling curve

processor &
hash key into the
processor's hash table

Figure 5.13. *A storage scheme for parallel adaptive methods: key-based addressing, a space-filling curve as hash function and a distributed hash table.*

to find the original node. This contrasts other procedures, where bookkeeping is necessary to find objects or their home processor.

The space-filling curve mapping can be considered as a hash function with a narrow distribution, because it introduces locality in the key addressing scheme, which is also exploited for the parallelisation of the code. Using the data locality once again on the local processor, one can optimise the usage of secondary disk storage and of the memory hierarchy of caches, which is difficult otherwise, see Douglas [107]. In order to increase the statistical variation of the hash function, we are using some simple scrambling of the binary representation of the hash key. However, some numerical experiments indicate that the overall performance of the hash table in fact does not really depend on the precise scrambling procedure. The space-filling curve mapping, i.e. the position on the curve can be computed for any given coordinate tuple. It is a unique mapping $[0,1]^d \to [0,1]$ similar to the mapping required for hash keys. A scrambled representation of the position can be also used as a hash key.

This framework for the parallelisation of adaptive codes by space-filling curves has originally been used for particle methods by Warren and Salmon [302]. Parallel hashing for adaptive mesh applications has been proposed even earlier by [308]. The space-filling curve partitioning has been generalised to programming environments for some grand challenge PDE projects by Parashar and Browne [236, 237]. Multigrid methods have been considered in Griebel and Zumbusch [151]. Similar parallelisation approaches are discussed in Schweitzer [271] and in Griebel et al. [141] for hierarchical particle algorithms.

In this chapter we have combined the different components which have been developed in the preceding chapters, namely multigrid methods, adaptive mesh refinement and mesh partitioning, in order to construct a parallel adaptive multigrid method. We have discussed the difficulties and possible methods to define multigrid methods on adaptively refined meshes, ranging from local multigrid to hierarchical and additive methods. Furthermore, we have reviewed parallel multigrid, where both the type of meshes and the algorithm play an important role. Parallel adaptive methods pose merely technical difficulties, which depend precisely on the algorithms on the adaptive meshes. Here often methods are used which cannot be extended easily to multigrid methods, so that the combination of all three components has been treated by a few codes only.

Our contribution in this field is the introduction of space-filling curves, which allow for a cheap and incremental parallel load balancing rather than more expensive graph algorithms. We are able to prove the efficiency of the partitioning itself and the multigrid algorithm on the partitions, which is unique. We do not expect a similar proof for general graphs by any competing partitioning method. Furthermore, space-filling curves allow for a deterministic partition, which in conjunction with key based addressing simplifies the parallel codes substantially. For the implementation we propose a parallel hash storage scheme, which fits nicely into the framework of key addressing and allows one to relate storage with space-filling curves for the improvement of memory cache effects.

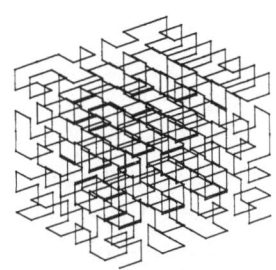

Chapter 6

Numerical Applications

In the previous chapter we have developed and proposed a parallel adaptive multigrid method. We were able to prove asymptotic parallel efficiency for mesh partitions created by space-filling curves. The proofs were based on the locality preserving properties of the curves. We also showed some numerical evidence that the cut sizes of the dual graphs of the meshes were indeed bounded. Now we want to verify the results for the whole parallel adaptive multigrid code experimentally. For this purpose we study several test problems on a variety of parallel computers. In order to demonstrate scalability of the algorithms, we use even some of the world's largest parallel computers, where we were able to run computations on up to 1024 processors.

We will present some numerical experiments to demonstrate the suitability and efficiency of the proposed space-filling curve load balancing. A parallel multigrid method on adaptively refined meshes with space-filling curve domain partitioning and a hash storage implementation is applied to a Poisson problem and the Lamé equation of linear elasticity. Furthermore, a parallel implementation of an adaptive sparse grid code again with space-filling curve domain partitioning and a hash storage implementation is used to solve a convection-diffusion and a ten-dimensional Poisson problem. The equations serve as simple test cases only. However, due to the few operations per node, it is rather hard to achieve good parallel efficiencies and, in this respect, these are even harder test problems than more complicated partial differential equations.

The numerical experiments were performed on three different large parallel computers, so that comparisons of a single application on several platforms are possible. The characteristics of the machines do influence the parallel performance and efficiency of an application. Especially the ratio of computing to communication power over a large scale of processors has a direct impact on the efficiency of a parallel code.

The computers are the PC cluster Parnass2 at our department with 72 dual Pentium II 400 MHz processors, a total of 144 processors, 512 Mbytes main memory per node and a Myrinet communication network (1.28 Gbit/s), configured as a fat-tree with a bisection bandwidth of 82 Gbit/s. For details see Griebel and Zumbusch [150] and Schweitzer, Zumbusch and Griebel [272]. The cluster was assembled and built at the department as the first 'self-made' or non-commercial installation outside of the US which ever entered the TOP 500 list of the world's largest computers. In the meantime, even larger clusters now exist in Europe, some of which are designed similarly. Performance data of the message passing implementations on Parnass2's Myrinet and alternative Fast Ethernet can be found in Table 6.1.

The simulations also have been carried out on ASCI Blue Pacific (called 'Blue'), which consists of several IBM SP computers. Each computing node has four PowerPC 604e processors at 332 MHz and a single network interface to a fat-tree network of IBM SP switches. The computer was built within DOE's ASCI initiative and is operated at Lawrence Livermore National Laboratory. Although the machine is said to have more than 5800 processors, it is partitioned into an open and several classified parts, so that a total of 1024 processors were available to unclassified research. The computer supports different MPI implementations, see [281]. One implementation of MPI is based on TCP/IP on the high performance network so that a maximum of 1024 processors can be addressed, which is coincidentally the maximum processor number available in the partition. A faster MPI implementation without this processor limit is based directly on the high performance network and transfers data through the user space memory. However, only a single process per node is able to communicate in this operation mode. The remaining three processors are either idle or can be utilised in a two-level parallelisation with multi-threading or automatic parallelisation. Performance numbers of both network modes are shown, four processors per node with MPI on TCP/IP (quadruple processor) and a single processor with MPI user space (uni processor). The raw message passing performance can be found again in Table 6.1. The successor of the computer, named ASCI white is again built by IBM, with eight-processor nodes Power3 332MHz and a new generation of SP switches. Unfortunately, at the time of writing the switches are not fully operational, so that ASCI Red and ASCI Blue Pacific were the world's largest computers.

Additional simulations have been done on some Cray T3E parallel computers. They are built with single processor nodes DEC/Compaq Alpha 21164 at speeds between 300 and 675 MHz, termed as T3E-600 to T3E-1350. The nodes

platform	MPI version	network	bandwidth [Mbit/s] theor.	max	∞	size $n_{max/2}$ [kbyte]	latency [μs]
SGI Indigo2	Mpich	Ethernet	10	7.9	7.9	2	1155
SGI O2	Mpich	Fast Eth.	100	45	45	2.5	625
Linux PC/Origin	Globus-Mpich	Fast Eth.	100	57	55	4.5	330
Linux PC	Mpich	Fast Eth.	100	89	89	2	105
Linux PC	Mpich	Gigabit Eth.	1000	378	320	3.5	75
Linux PC	Mpich TCP/IP	memory	6400	215	180	6	138
Linux PC	Mpich shmem	memory	12800	3000	1000	10	14
DEC Cluster	DEC MPI	Mem. Chan.	800	450	350	2.5	15
DEC Cluster	Mpich MPI	memory	8800	626	390	1	8.3
DEC Cluster	DEC MPI	memory	8800	774	500	1	4.2
ASCI Blue Pac.	TCP/IP	IBM SP	1200	240	240	80	326
ASCI Blue Pac.	TCP/IP	memory	10640	207	207	30	247
ASCI Blue Pac.	user space	IBM SP	1200	675	675	17	34
ASCI Blue Pac.	user space	memory	10640	530	530	22	49
ASCI Blue Pac.	shmem	memory	10640	775	770	3.5	17
Linux PC	Scampi	SCI	4000	680	680	2.5	9
Linux PC	Scampi	memory	6400	1435	615	1	4
Parnass2	HPVM	Myrinet	1280	780	530	0.7	5.3
Parnass2	HPVM	memory	6400	810	625	6	14
Parnass2	SCore	Myrinet	1280	825	810	1	12
Parnass2	SCore	memory	6400	1176	700	2	10
SGI Origin 2000	SGI MPI	memory	6400	1034	1034	3.5	16
Cray T3E-900	Cray MPI	Cray T3E	4000	1305	1305	9	23
Cray T3E-1200	Cray MPI	Cray T3E	4000	1420	1420	9	21
ASCI Red	Intel MPI	Intel Paragon	6400	2603	1280	8	20

Table 6.1. *Performance of MPI message passing libraries on several parallel computers: Network bandwidth, message size $n_{max/2}$ for half of the sustained bandwith and message latency.*

are connected by a three-dimensional torus network directly attached to the memory controller and the 3rd level cache interface. The performance of the MPI implementation is depicted in Table 6.1. However, even more impressive rates can be obtained with a shared memory programming paradigm. Several of these machines are installed world wide. We report on numbers obtained on two 512 processor Cray T3Es located at NIC Forschungszentrum Jülich.

However, even larger machines exist, with all machines with 1024 processors and above being classified and at US agencies. Our 1024 processor numbers were measured at Cray Inc. at the time when one of the later classified machines was assembled and tested. It was a 600 MHz T3E-1200 with roughly 1400 processors, the largest computer Cray ever built.

6.1 Parallel Multigrid for a Poisson Problem

First of all, we consider the Poisson equation

$$-\Delta u \;=\; f \quad \text{in } \Omega = [-1,1]^d, \; d = 2,3$$

with the Dirichlet boundary conditions $u = 0$ on part of the boundary $\Gamma_D \subset \partial\Omega$ and homogeneous Neumann boundary conditions $\frac{\partial}{\partial n} u = 0$ on the remaining boundary $\Gamma_N = \partial\Omega \setminus \Gamma_D$. In the following, we consider two types of boundary conditions Γ_D. First, we consider a pure Dirichlet problem $\Gamma_N = \emptyset$, which results in a smooth solution. Here, a uniformly refined mesh is sufficient for discretisation and adaptivity is not needed. Second, we consider the case where the solution possesses almost a corner singularity due to the source term f. We run our adaptive multilevel finite difference code to solve it.

A nested iteration is used for a solution up to discretisation error. With an additive multilevel V-cycle preconditioned Krylov iteration on each mesh this means a constant number of iterations. Together with a geometrically increasing number of mesh points we end up with a total of $\mathcal{O}(n)$ operations. The adaptive mesh refinement is controlled by a residual based error indicator, leading to a parallel adaptive mesh refinement of one-irregular mesh built of hexahedral elements, where a finite-difference-type discretisation is employed. The parallel version of the code uses a space-filling curve data partition with Hilbert curves after each adaptive mesh refinement step.

6.1.1 Uniformly Refined Meshes

In the first test we consider regular meshes and uniform mesh refinement. Tables 6.2 and 6.3 show wall clock times for the solution of the equation system on a regular mesh of different levels within the nested iteration using different numbers of processors from level 4 to 11 on Parnass2 and Blue Pacific.

Analogous three dimensional simulations are shown in Tables 6.4 and 6.5 on Parnass2 and Blue Pacific from level 3 to 7. Some of the numbers are also compared graphically in Figure 6.1.

For the uniform refined mesh cases we observe scalability of the algorithms in n and p in general with different parallel overhead and processor speed.

time		processors						
level	nodes	1	2	4	8	16	32	64
4	289	0.44	0.37	0.27	0.27			
5	1089	3.30	2.18	1.32	0.94	0.69	0.52	0.43
6	4225	16.0	8.52	4.70	2.88	1.74	1.16	0.79
7	16641	66.9	33.6	17.6	9.87	5.40	3.17	1.88
8	66049	283	141	70.4	36.2	19.0	10.4	5.64
9	263169	1160	583	291	147	73.8	37.5	19.4
10	1050625			1162	584	294	147	74.0
11	4198401					1162	585	292

Table 6.2. *Poisson problem, uniformly refined meshes in two dimensions, timing of the multigrid solution on Parnass2.*

time		processors						
level	nodes	1	4	4 uni proc	16	64	256	512
4	289	0.81	1.68	1.42	2.31	4.50	10.67	11.36
5	1089	9.51	4.02	3.74	3.40	5.35	10.66	24.79
6	4225	45.17	12.97	12.48	6.90	6.76	12.45	34.35
7	16641	267.7	53.08	51.77	18.28	14.15	18.15	23.59
8	66049	1722	302.4	289.3	72.92	29.20	33.10	32.58
9	263169		1891	1792	419.0	95.47	58.31	67.26

Table 6.3. *Poisson problem, uniformly refined meshes in two dimensions, timing of the multigrid solution on ASCI Blue Pacific.*

time		processors						
level	nodes	1	2	4	8	16	32	64
3	729	3.05	2.07	1.51	1.15	1.14	1.01	
4	4913	30.2	17.6	10.6	6.87	5.11	3.64	2.73
5	35937	277	150	81.4	46.4	29.0	17.9	11.4
6	274625	2455	1297	674	356	198	109	61.8
7	2146689				2782	1482	774	410

Table 6.4. *Poisson problem, uniformly refined meshes in three dimensions, timing of the multigrid solution on Parnass2.*

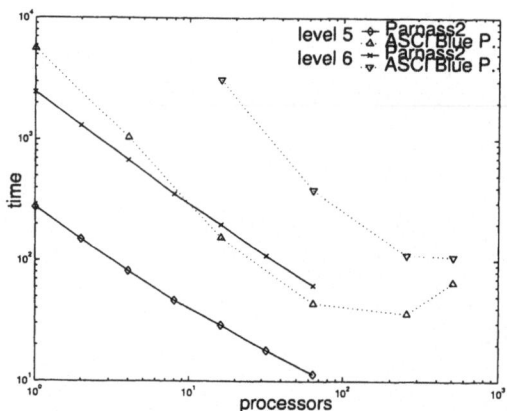

Figure 6.1. *Poisson problem, uniformly refined meshes in three dimensions, timing of the multigrid solution on level 5 and 6 on Parnass2 and ASCI Blue Pacific.*

time					processors			
level	nodes	1	4	4 uni proc	16	64	256	512
3	729	7.41	5.55	4.69	4.64	8.26	17.34	26.51
4	4913	142.8	55.85	53.65	17.19	12.63	20.65	33.76
5	35937	5617	1088	1036	155.1	44.01	36.18	65.32
6	274625				3116	383.4	110.5	106.6

Table 6.5. *Poisson problem, uniformly refined meshes in three dimensions, timing of the multigrid solution on ASCI Blue Pacific.*

The number of degrees of freedom and the amount of work increases roughly by a factor of 2^d from one row to the next, whereas the number of processors increases by factors of two or four from column to column. Hence we expect increasing computing times from row to row by a factor of 2^d and ideally decreasing times from column to column by factors of $1/2$ or $1/4$ respectively. Furthermore we can compare the absolute times of the different platforms. In Figure 6.1 we also see selected rows of the three dimensional examples in a log-log scale, number of processors versus time. Straight lines with a slope of 1 : 1 would indicate an ideal scale up. While speedup defines the acceleration of a fixed problem size by p processors, which would ideally be p, the term scaleup means timing of a problem with an increase in the number of operations and an increase of the number of processors by the same factor. The scaleup should come close to one, as does the parallel efficiency, which is simply the speedup in relation to the number of processors p. The efficiency for large numbers of processors will often be lower than the scaleup, because there may not be enough work left per processor for the parallel efficiency, whereas the amount of work per processor is fixed for the scaleup.

For a fixed number of processors, we observe a scaling of a factor of 2^d from one level to the next finer level which corresponds to the factor of 2^d increase in the amount of unknowns on that level. Furthermore, for a fixed level the measured times scale roughly with $1/p$ of the number of processors. However, the 32 and 64 processors on Parnass2 and the 64 and 256 processors on Blue Pacific perform efficiently only for sufficiently large problems, i.e. for problems with more than some thousand degrees of freedom per processor. The 512 processor cases on Blue Pacific seem to be slower in general, even for large problem sizes. This may indicate that even larger problem sizes are needed and that the communication network of Blue Pacific does not scale that well.

There are two different four processor numbers on Blue Pacific. The first column is measured on a single four processor node with MPI on IP, but without the network. The second column 'uni proc' uses four nodes with four processors each, where only a single processor is used per node. In this case the network interface can be used by a single process, and communication over the network is faster than the usual shared network devices. Of course this is a waste of processors. Hence we did not run larger configurations in single processor mode for this test case. However, we see that this mode is slightly faster than the four-processor mode, which may also indicate a lack of memory bandwidth of the four processor SMP nodes. Prototype nodes of ASCI white show an even more pronounced lack of memory bandwidth, since eight processors share a single memory subsystem there.

For large problems on Parnass2 and Blue Pacific we can even obtain some super-linear speedups, probably due to caching effects. If we fix the amount of work, that is the number of nodes per processor, we will obtain the scaleup. Comparing a time at one level l and a number of processors p with the time of one level finer $l + 1$ and $2^d \cdot p$ processors, we obtain a nearly perfect scaling of the method in large parameter domains. The two-dimensional example shows slightly higher parallel efficiencies than the tree-dimensional example, because the data to be exchanged for n degrees of freedom increases from $\mathcal{O}(n^{1/2})$ (2D) to $\mathcal{O}(n^{2/3})$ (3D), which makes three dimensional problems harder to parallelise.

Note that in this case of uniform mesh refinement, an a priori partition of the uniform meshes into stripes or square shaped cells would be superior to any dynamic partitioning scheme. However, our space-filling curve load balancing

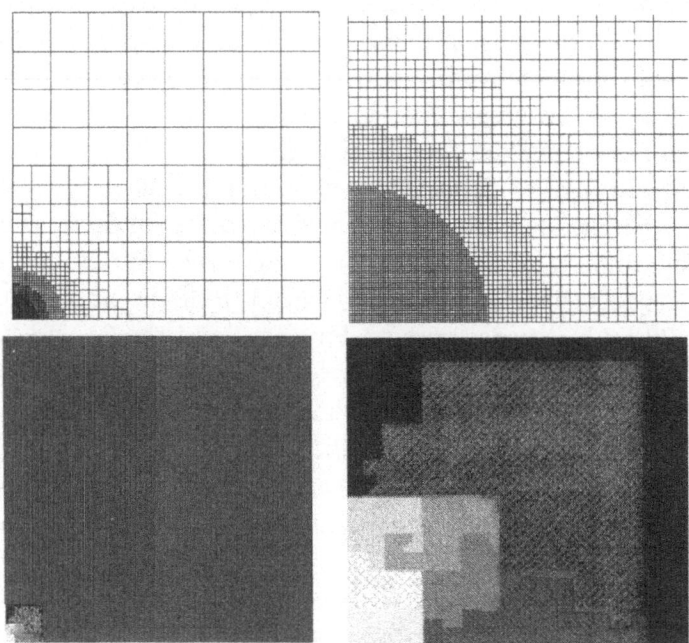

Figure 6.2. *Poisson problem, adaptively refined meshes in two dimensions mapped to 8 processors. Full domain (left), zoomed image (right), mesh and grey coded partitions.*

scheme performs well and introduces little overhead, which can also be seen in the next section 6.1.2.

6.1.2 Adaptively Refined Meshes

In the next test we consider adaptively refined meshes for a problem with singularities, where the meshes are refined toward a singularity located in the lower left corner, see also Figure 6.2. Here partitions are depicted which are obtained by Hilbert space-filling curve partitioning.

Tables 6.6 and 6.7 show times in the two-dimensional adaptive case on Parnass2 and the Cray T3E with 300MHz processors. These numbers give the wall clock times for the solution of the equation system again within the nested iteration, now on different levels of adaptive meshes and on different numbers of processors.

Similar numbers for the three-dimensional case can be found in Tables 6.8 and Table 6.9 on Parnass2 and again a Cray with 300MHz processors. We list

time nodes	processors						
	1	2	4	8	16	32	64
134	0.37	0.26	0.24	0.24	0.24	0.27	0.27
224	0.69	0.49	0.36	0.34	0.31	0.30	0.32
384	1.27	0.85	0.69	0.51	0.42	0.37	0.35
682	2.38	1.48	1.04	0.75	0.57	0.45	0.41
1243	4.54	2.81	1.81	1.21	0.83	0.60	0.51
2320	8.75	4.92	3.13	1.95	1.25	0.86	0.62
4391	17.0	9.30	5.19	3.26	1.89	1.25	0.85
8460	33.5	17.8	10.1	5.57	3.27	1.92	1.26
16469	66.9	34.4	18.1	10.1	5.50	3.21	1.99
32291	133	67.7	35.4	19.3	10.3	5.50	3.27
63736	263	134	68.5	36.6	19.2	10.5	5.66
126271	529	272	139	70.4	36.7	19.1	10.3
250911		560	278	143	71.8	36.8	19.1

Table 6.6. *Poisson problem, adaptively refined meshes in two dimensions, timing of the multigrid solution on Parnass2.*

time nodes	processors					
	1	4	16	64	128	256
1089	5.08	1.27	0.72	0.64	0.84	1.30
1662	5.85	2.01	0.97	0.72	0.86	1.33
2745	10.7	3.26	1.37	0.85	0.94	1.38
4834	20.3	5.84	2.01	1.08	1.08	1.46
8915	39.8	10.9	3.38	1.42	1.26	1.56
16948	78.5	39.7	5.68	2.08	1.66	1.78
32788	157	77.7	10.7	3.34	2.30	2.14
64251			20.7	5.97	3.62	2.80
126810				10.9	6.14	4.12
251468					11.2	6.64
500135					21.2	11.7
996531					41.0	21.8
1988043					80.6	41.4

Table 6.7. *Poisson problem, adaptively refined meshes in two dimensions, timing of the multigrid solution on a Cray T3E-600.*

time nodes	processors						
	1	2	4	8	16	32	64
1191	5.82	3.64	2.53	1.75	1.35	1.21	1.22
2178	12.3	7.94	5.07	3.82	3.02	2.20	1.97
4454	28.5	17.0	11.0	7.00	4.74	3.57	3.00
10061	71.9	43.9	26.5	16.2	10.2	7.04	5.16
24215	190	108	60.8	36.3	21.0	14.0	9.07
61361	510	280	157	87.7	49.0	29.5	17.7
160384	1418	772	404	217	125	70.5	40.8
429613				602	318	175	95.8

Table 6.8. *Poisson problem, adaptively refined meshes in three dimensions, timing of the multigrid solution on Parnass2.*

time nodes	processors					
	1	4	16	64	128	256
35937	291	85.6	29.6	11.2	7.61	5.94
50904	423	129	41.0	14.8	10.1	7.17
89076	405	236	71.2	24.6	14.6	9.98
189581			154	49.7	29.2	17.2
460421				109	61.1	35.6
1201650					142	77.2
3251102					345	188

Table 6.9. *Poisson problem, adaptively refined meshes in three dimensions, timing of the multigrid solution on a Cray T3E-600.*

execution times of the linear equation solver for different problem sizes and numbers of processors.

Note that it is more difficult to compare these numbers, since we have been using different initial coarse meshes. Furthermore, the adaptive mesh refinement is sensitive to small changes in the solution, so that even small differences in rounding may cause different mesh refinement. Although the methods converge to the same solution, the discrete meshes on which the solution is represented may differ. In order to compare the given execution times, the different number of degrees of freedom have to be taken into account. Furthermore, the mesh structure and the number of mesh levels involved may also differ.

ratio α	nodes	processors						
		1	2	4	8	16	32	64
uniform 2D	66049	0	9.7e-4	1.1e-3	1.3e-3	1.3e-3	3.2e-3	4.9e-3
adaptive 2D	63736	0	6.8e-4	7.9e-4	9.1e-4	1.1e-3	2.9e-3	3.5e-3
uniform 3D	35937	0	3.2e-4	4.1e-4	1.9e-3	1.4e-3	1.6e-3	3.7e-3
adaptive 3D	61361	0	3.4e-4	7.2e-4	8.8e-4	9.8e-4	1.1e-3	1.5e-3

Table 6.10. *Ratio α of execution times of partitioning and mapping nodes to solving the equation system, Poisson problem on Parnass2.*

We obtain a scaling of about a factor two to three from one level to the next finer level, i.e. the times are proportional to the number of unknowns for a fixed number of processors. This is due to the adaptive mesh refinement heuristic. Increasing the number of processors speeds up the computation accordingly. We get speedup also for large numbers of processors up to 256, as long as the problem sizes are large enough. The direct comparison of the times on the Cray and on Parnass2 show slight advantages for Parnass2. This does not represent the theoretical peak performance of the processors but can be explained by the access times of the memory system, the superior integer processing performance of the Pentium processors and the quality of the compilers.

Now we compare the time for solving the equation system with the time required for sorting the nodes according to the space-filling curve and mapping them to processors before the computation starts. We compare the execution time of the load balancing step with the execution time of the linear solver and compute their ratio $\alpha := t_{\text{balancing}}/t_{\text{solving}}$. Table 6.10 gives the ratios partitioning to solving on Parnass2 for the examples of the previous uniform and the current adaptive mesh refinement in two and three dimensions using different numbers of processors.

In the single processor case, no load balancing is needed, so the partitioning and mapping time and the ratio α is zero. Otherwise, the nodes have to be partitioned and mapped by a parallel (partial) sorting algorithm. In the uniform mesh case the relative cost of partitioning nodes α is of the order $1e - 3$. In the case of uniform refinement, for a refined mesh, there are only few nodes located at processor boundaries which may have to be moved during the mapping. Hence our load balancing is also very cheap in this case. In the adaptive mesh case, dynamic load balancing is generally required. Note that in all cases load balancing is much cheaper than solving the equation system.

However, higher numbers of processors make the mapping and partitioning slightly slower. Mapping data for adaptive refinement requires the movement of a large amount of data, even if most of the nodes stay on the processor. Other load balancing strategies can be quite expensive for adaptive refinement procedures and lead to different partition quality, see Bastian [26] and Stals [285].

6.2 Parallel Multigrid for Linear Elasticity

As another test case for our approach we consider the linear elasticity problem. The Lamé equation (see Braess [54] or Ciarlet [89]) in the displacement formulation, either as a linear three dimensional problem or as a two dimensional plain-strain problem, reads as

$$
\begin{aligned}
\mu \vec{\Delta} \vec{u} + (\lambda + \mu)\nabla(\nabla \cdot \vec{u}) &= \vec{f} && \text{in } \Omega \subset \mathbb{R}^d,\ d = 2,3 \\
\vec{u} &= \vec{0} && \text{on } \Gamma_D \subset \partial\Omega \\
\sigma(\vec{u}) \cdot \vec{n} &= \vec{0} && \text{on } \Gamma_N = \partial\Omega \setminus \Gamma_0
\end{aligned}
\tag{6.1}
$$

with Lamé's constants

$$
\lambda = \frac{E\nu}{(1+\nu)(1-2\nu)} \qquad\qquad \mu = \frac{E}{2(1+\nu)}
$$

and linearised strain $\epsilon(\vec{u}) = \frac{1}{2}(\frac{\partial u_i}{\partial x_j} + \frac{\partial u_j}{\partial x_i})_{ij}$, $\lambda > 0$, $\mu > 0$ and $E > 0$. The problem is elliptic due to Korn's inequality for $0 < \nu < \frac{1}{2}$ (see e.g. Braess [54]) and can be solved by a multilevel preconditioned iterative Krylov solver. Although this is a system of equations, for moderate Poisson ratio $\nu < \frac{1}{2}$ it is sufficient to use the scalar multilevel method and to apply it to each component u_i. Larger incompressibility of the material ν would require a different multigrid strategy, for example a block-smoother with $d \times d$ blocks at each nodal point. Again we use a Hilbert curve for the geometric load balancing of the meshes, where nodes (with d degrees of freedom) are mapped to the processors.

6.2.1 Uniformly Refined Meshes

We consider two test cases: first, we take a homogeneous square or cube under internal forces parallel to one coordinate axis, which is fixed at two adjacent edges ($d = 2$) or three faces ($d = 3$). The remaining edges or faces are free and unloaded (homogeneous Neumann conditions). The solution is smooth and therefore uniform refined meshes will be sufficient. Wall clock times of the

	time		processors						
level	nodes	dof	1	2	4	8	16	32	64
4	289	578	1.27	0.74	0.48	0.34	0.26	0.22	0.22
5	1089	2178	5.31	2.83	1.60	0.96	0.61	0.41	0.34
6	4225	8450	21.9	11.2	5.93	3.24	1.77	1.05	0.69
7	16641	33282	91.6	45.3	23.1	12.0	6.19	3.35	1.89
8	66049	132098	386	192	94.2	46.7	23.6	12.2	6.34
9	263169	526338	1578	789	392	195	95.8	47.3	23.9
10	1050625	2101250			1559	777	388	192	95.6

Table 6.11. *Linear elasticity problem, uniformly refined meshes in two dimensions, timing of the multigrid solution on Parnass2.*

	time		processors						
level	nodes	dof	1	2	4	8	16	32	64
3	729	2187	3.05	1.99	1.32	0.91	0.73	0.53	
4	4913	14739	27.6	15.3	8.81	5.49	3.51	2.07	1.30
5	35937	107811	249	131	68.2	38.1	21.4	11.7	6.57
6	274625	823875	2164	1124	573	300	158	80.5	41.7
7	2146689	6440067					1211	618	311

Table 6.12. *Linear elasticity problem, uniformly refined meshes in three dimensions, timing of the multigrid solution on Parnass2.*

iterative solver for the two-dimensional and three-dimensional case obtained on Parnass2 are shown in Tables 6.11 and 6.12.

In the case of linear elasticity we have d-times more degrees of freedom than nodes. Furthermore, the number of iterations for the solution of the equation system is higher than for the scalar case. Nevertheless, in the relative comparison of different numbers of processors or different problems sizes, these differences are cancelled. We obtain almost linear speedups for large problem sizes. Furthermore, scaleups can be read easily in the table of the two dimensional problem from the comparison of a number and the corresponding entry two columns left and one row down. We see very good scaleups for the two dimensional case and almost as good numbers for the three dimensional case, which again demonstrates that the three-dimensional case is harder to parallelise.

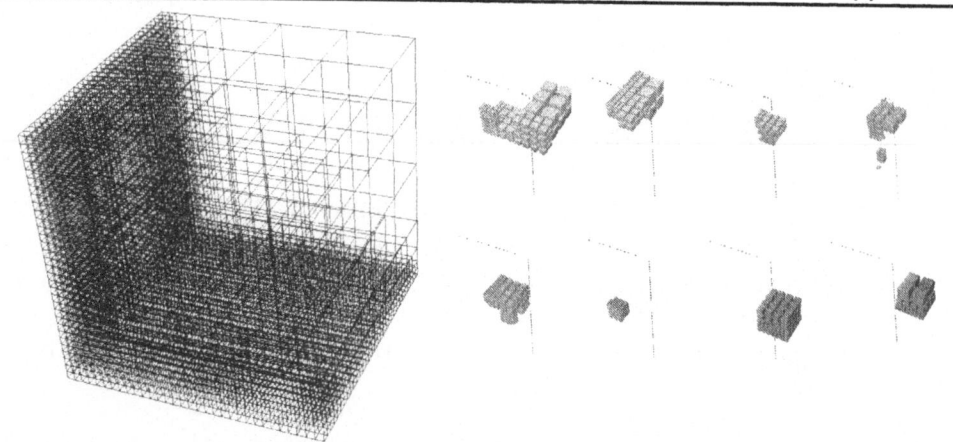

Figure 6.3. *Linear elasticity problem, adaptively refined meshes in three dimensions mapped to 8 processors. Mesh (left), grey shaded partitions on the (right).*

6.2.2 Adaptively Refined Meshes

As a second test problem, we consider a homogeneous cube under internal gravitational forces parallel to one coordinate axis which is fixed at three faces. The remaining edges or faces are free and unloaded (homogeneous Neumann conditions). The solution develops singularities due to the boundary conditions and driven by the volume forces of the right hand side. Here adaptive refinement is used to resolve the singularities next to the edges or faces (left, bottom and front) which separate homogeneous Dirichlet (fixed faces) and homogeneous Neumann conditions (free faces), see Figure 6.3 (left). All numbers reported are again CPU times measured on the different parallel computers.

For this experimental setup, we are able to compare a larger number of computers and configurations. We present wall clock times on a 1024 processor Cray T3E-1200 (Table 6.13), on ASCI Blue Pacific with up to 1024 processors (Table 6.14) and on Parnass2 (Table 6.15). Furthermore, we have performed the same experiments on alternative configurations of Parnass2, with a single processor per communication interface instead of two processors and with 100 Mbit/s Fast Ethernet instead of Myrinet 1280 Mbit/s, see Tables 6.16, 6.17, 6.18. Thus we are able to vary the ratio of computation to communication, and we can study the influence of this ratio on the parallel performance of the code. Part of the numbers are also drawn in Figure 6.4 for a short overview.

One can see principally also the same behaviour of our approach for Lamé's

time		processors								
nodes	dof	1	4	16	64	128	256	512	768	1024
35937	107811	162	34.1	9.11	2.23	1.23	0.75	0.55	0.56	0.52
109873	329619	435	108	29.6	7.20	3.57	1.87	1.13	0.91	0.80
410546	1231638			114	28.6	14.2	7.02	3.51	2.48	1.94
1857030	5571090				133	67.1	33.3	16.5	11.0	
9619175	28857525					351				

Table 6.13. *Linear elasticity problem, adaptively refined meshes in three dimensions, timing of the multigrid solution on a Cray T3E-1200.*

equation as for the Poisson problem. Note that the parallel efficiency is even higher than for the Poisson problem. This is due to the higher amount of work associated with each node, while the expenses associated with the mesh stay the same. In the elasticity case, there are d degrees of freedom located at each node.

In general, the numbers show that the presented algorithm scales very well. We obtain a scaling of about a factor 3–5 from one level to the next finer level, i.e. the times are proportional to the number of unknowns for a fixed number of processors. This is due to the adaptive mesh refinement heuristic. For a fixed number of processors, we observe a scaling from one level to the next finer level by a factor which corresponds to the increase in the number of unknowns on that level. Furthermore, for a fixed level the measured times scale roughly with $1/p$ in the number of processors. However, the 64 processors and more on Parnass2 perform efficiently only for sufficiently large problems, i.e. for problems with more than some ten thousand degrees of freedom. The behaviour on the Cray and on Blue Pacific is similar, in that a parallel efficient use on the Cray of 256 processors and on ASCI Blue Pacific of 64 processors and above requires problems that are large enough. For larger problems we even obtain some super-linear speedups, probably due to caching effects. If we fix the amount of work, that is the number of nodes per processor, we will obtain the scaleup. Comparing a time at one level l and a number of processors p with the time of one level finer $l + 1$ and roughly $4 \cdot p$ processors, we obtain a nearly perfect scaling of the method.

A direct comparison of the execution times reveals that the 300 MHz Alpha processors in the Cray T3E-600 configuration perform roughly a factor 2.5 slower than the Parnass2 nodes and the 600 MHz Alpha processors, and the Power PC 604e processors of Blue Pacific are faster than the Pentium II processors of Parnass2. The highly sophisticated T3E network scales slightly

time	processors					
	1	4	16	64	256	1024
uni processor	36.5	37.1	8.67	17.15		
quadruple processor	36.5	4.31	1.22	2.67	42.4	17.5

Table 6.14. *Linear elasticity problem, adaptively refined meshes in three dimensions, 35937 nodes and 109873 degrees of freedom, timing of the multigrid solution on ASCI Blue Pacific.*

time		processors						
nodes	dof	1	2	4	8	16	32	64
35937	107811	10.75	5.42	2.70	1.42	0.76	0.39	0.20
109873	329619	367.0	190.2	95.77	51.13	25.64	12.85	6.51
410546	1231638	1438	758.3	367.9	202.2	100.8	51.54	24.91
1857030	5571090			1724	921.9	467.6	239.7	117.1

Table 6.15. *Linear elasticity problem, adaptively refined meshes in three dimensions, timing of the multigrid solution on Parnass2, dual-processors connected by Myrinet 1280 Mbit/s.*

better than the Parnass2 Myrinet network.

ASCI Blue Pacific shows some kind of erratic timing behaviour for large numbers of processors which is also supported by large standard deviation of the measured timings. In addition to obvious difficulties with the uni-processor timings, where only one of four processors on each node is used, the quadruple processor configuration also shows an early saturation. This is in contrast to the timings we have seen so far on Blue Pacific for uniformly refined meshes. The larger amount of data to transfer and the more irregular communication patterns due to the space-filling curve partitioning in the adaptive mesh case seem to interfere with the communication infrastructure of Blue Pacific, which also seems to suffer from side effects of other parallel jobs that were run concurrently on other partitions of the machine.

The set of data on Parnass2 with different network interfaces and node configurations allows a more careful analysis than the data gathered on Blue Pacific. The variation of the available network bandwidth ranges from effective 45 Mbit/s for the double processor Fast Ethernet case to 820 Mbit/s for the single processor Myrinet case and well above these numbers for the Cray T3E, see also the detailed analysis in Table 6.1. The CPU times for a fixed

time		processors					
nodes	dof	1	2	4	8	16	32
35937	107811	10.75	5.33	2.64	1.40	0.75	0.38
109873	329619	367.0	184.7	92.41	49.83	25.30	12.64
410546	1231638	1438	730.1	360.2	198.5	99.39	49.18
1857030	5571090		3351	1692	912.3	460.3	228.7

Table 6.16. *Linear elasticity problem, adaptively refined meshes in three dimensions, timing of the multigrid solution on Parnass2, uni-processors connected by Myrinet 1280 Mbit/s.*

time		processors							
nodes	dof	1	2	4	8	16	32	48	64
35937	107811	10.80	5.42	3.55	1.88	1.58	0.94	0.60	0.51
109873	329619	367.9	187.3	116.3	108.8	53.85	44.11	35.90	33.41
410546	1231638	1443	795.8	447.0	350.1	213.0	116.0	75.27	59.41
1857030	5571090						451.5	297.4	249.61

Table 6.17. *Linear elasticity problem, adaptively refined meshes in three dimensions, timing of the multigrid solution on Parnass2, dual-processors connected by Fast Ethernet 100 Mbit/s.*

problem size are also assembled in Figure 6.4. We clearly see that the double processor Fast Ethernet case deviates from the almost linear scaleup behaviour of the other test cases at eight processors and higher. Hence this amount of network bandwidth seems to be inappropriate for the demands of the parallel algorithm. The effect that the single processor configuration with the same network interface performs much better may be due to an effective load sharing on the double processor nodes, where one processor runs the parallel code and the second processor handles the networking. A similar behaviour can also be observed on ASCI Red with the Intel double processor boards, which in that case leads to a brilliant effective network performance. Nevertheless, there are slight advantages of the faster Myrinet compared to the Fast Ethernet network simulations for the single processor case and a dramatic improvement for the double processor test cases. Again we see that the network performance of the Cray T3E architecture is highly scalable and demonstrates good speedups even up to 1024 processors. Small deviations from the optimum straight line are due to the limited problem size, which can also be seen in the more detailed

| time | | processors | | | | | | |
nodes	dof	1	2	4	8	16	24	32
35937	107811	10.80	5.37	2.70	1.48	0.81	0.57	0.45
109873	329619	367.9	185.1	95.06	51.56	26.59	18.36	14.06
410546	1231638	1443			203.6	102.8	68.89	52.42
1857030	5571090					472.6	320.1	242.95

Table 6.18. *Linear elasticity problem, adaptively refined meshes in three dimensions, timing of the multigrid solution on Parnass2, uni-processors connected by Fast Ethernet 100 Mbit/s.*

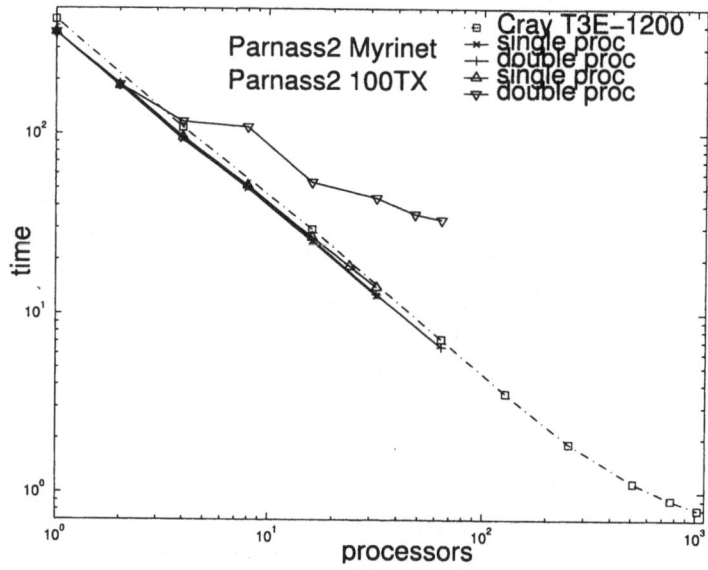

Figure 6.4. *Linear elasticity problem, adaptively refined meshes in three dimensions, timing of the multigrid solution on a Cray T3E-1200 and on Parnass2 with Myrinet (1280 Mbit/s) and Fast Ethernet (100 Mbit/s) interconnection network and different node configurations.*

Table 6.13.

The previous experiments were based on a rather fine initial mesh to allow scaling up to 1024 processors. Now we choose a much coarser initial mesh and start the adaptive mesh refinement. Consequently, the sequence of meshes differs. Therefore, the number of unknowns per level and the corresponding amount of work for the multilevel V-cycle are not the same in the experiments.

| time | | processors | | | | | | |
nodes	dof	1	2	4	8	16	32	64
125	375	0.10	0.12	0.11	1.50	2.91		
450	1350	1.44	0.99	0.80	1.35	0.50	0.39	1.05
1155	3465	4.14	2.48	1.71	1.32	1.00	0.70	2.74
4412	13236	19.0	10.3	6.09	5.23	3.07	1.89	1.21
18890	56670	98.6	50.3	28.1	20.6	11.6	6.35	3.70
93021	279063	582	294	157	102	54.8	28.2	15.1
506620	1519860				556	306	155	78.1
3178218	9534654							494

Table 6.19. *Linear elasticity problem, adaptively refined meshes in three dimensions, starting from a different initial coarse mesh, timing of the multigrid solution on Parnass2.*

| ratio α | | processors | | | | | | |
	nodes	1	2	4	8	16	32	64
uniform 3D	274625	0	3.4e-4	3.2e-4	3.3e-4	1.7e-3	6.7e-4	1.4e-3
adaptive 3D	18890	0	3.5e-4	3.4e-4	3.9e-4	6.0e-4	1.15e-3	2.73e-3
adaptive 3D	93021	0	3.6e-4	3.5e-4	4.0e-4	4.9e-4	6.9e-4	1.43e-3
adaptive 3D	506620	0	—	—	4.2e-4	4.6e-4	6.3e-4	9.1e-4

Table 6.20. *Ratio α of execution times of partitioning and mapping nodes to solving the equation system for a linear elasticity problem, adaptively refined meshes in three dimensions, timing of the multigrid solution on Parnass2.*

Table 6.19 shows numbers obtained on Parnass2.

Further data taken from the experiments, see Table 6.20, reveal that the effort and computing time spent on load balancing is below 1% of the time needed for the solution of the linear equation systems and is hence negligible. We actually can afford to perform a load balancing every time that an adaptive mesh refinement takes place and the mesh changes. In other words, the computation is load balanced at every step of the overall computation.

6.3 Parallel Solvers for Sparse Grid Discretisations

In the previous sections we saw a number of numerical experiments of parallel multigrid methods for standard discretisations. Uniform and adaptively refined

nodes	1	4	4 uni proc	16	64	256	512
59049	210.6	89.07	85.97	33.91	12.13	20.95	132.3
452709	2989	1462	1385	556.6	198.1	63.13	37.77
2421009					2290	768.5	458.9

Table 6.21. *Poisson problem, uniformly refined sparse grids in ten dimensions, timing of the iterative solver on ASCI Blue Pacific.*

meshes, the Poisson equation and the Lamé equation were considered on a variety of parallel computers. We conclude the chapter with numerical results on a different type of discretisation, namely sparse grids. We have a look at two different test cases, a ten-dimensional Poisson equation, which is something that is clearly not possible for standard discretisations. The second example is a three-dimensional convection-diffusion equation on adaptively refined sparse grids.

6.3.1 Uniformly Refined Sparse Grids

As a first test we present a ten-dimensional Poisson problem discretised using finite differences on a uniform sparse grid. It is partitioned by a space-filling curve heuristic with domain-slicing. In Table 6.21 we present the execution times on ASCI Blue Pacific with up to 512 processors.

Similar to the cases of parallel multigrid on uniformly refined meshes, we obtain some good speedups on ASCI Blue Pacific up to 512 processors, with the exception of the smallest test case of the table, which scales only up to 64 processors. However, the absolute speedups are not as large as for the multigrid case. This can be explained by the amount of data to be transferred within the sparse grid discretisation. As we saw in chapter 4.4 on grid partitioning of sparse grids, we cannot expect a scaling as brilliant as for standard local discretisations, as long as we do not consider autonomously large problems. Instead of a $n^{1/d}$ ratio of computing to communication we observe only a $\log n$ term, which causes the difficulties. Nevertheless, the numerical simulations can be considered as efficient in the sense that the parallelisation was able to speed up this ten-dimensional simulation, which would not be possible for standard type of discretisations. Note that an even more efficient parallelisation would have been possible for uniformly refined sparse grids by a combination technique type of parallelisation, where completely concurrent jobs would have been created. However, the space-filling curve partitioning

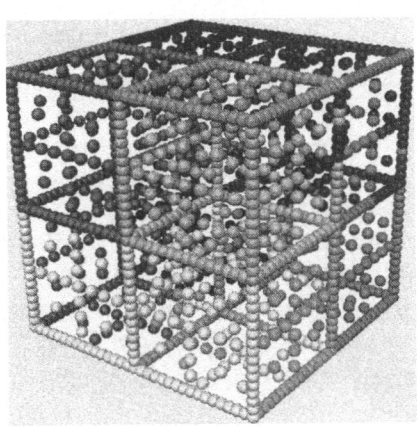

Figure 6.5. *A three-dimensional sparse grid, partitioned to eight processors (grey coded).*

proved effective also in this case.

As a second test case we present a three-dimensional convection-diffusion problem discretised using finite differences on a uniformly refined and on an adaptively refined sparse grid. A partition of a three-dimensional uniformly refined sparse grid is depicted in Figure 6.5. The wall clock times for the solution of the equation system on a sparse grid of different levels using different numbers of processors are shown for Parnass2 in Table 6.22, for a Cray T3E-600 in Table 6.23 and for ASCI Blue Pacific in Table 6.24. Part of the numbers are also assembled in Figure 6.6 for comparison of the three parallel computers.

For a fixed number of processors, we observe a scaling by a factor slightly above of 2 from one level to the next finer level, which corresponds to a similar factor of increase in the amount of unknowns on that level. Furthermore, for a fixed level the measured times scale roughly with $1/p$ for the number of processors p up to a parallel efficiency of 40% for 64 processors on Parnass2. However, the 32 and 64 processors on Parnass2 perform efficiently only for sufficiently large problems, i.e. for problems with more than some thousand degrees of freedom. The parallel efficiencies for 256 processors drop to 24% on the Cray and to 16% for the same test case of ten refinement levels. Again the problem size has to scale along with the number of processors for efficiency reasons. The absolute execution times show that in this example the 300MHz Cray is a constant factor slower than Parnass2, whereas Blue Pacific operates roughly at the same speed as Parnass2. Again this does not reflect the theoretical peak performance of the processors, which would definitely be in favour

level	time nodes	1/h	processors						
			1	2	4	8	16	32	64
2	81	4	0.03	0.03	0.07	0.08	0.11		
3	225	8	0.12	0.09	0.09	0.11	0.16	0.20	0.20
4	593	16	0.63	0.41	0.33	0.32	0.38	0.44	0.53
5	1505	32	3.78	2.29	1.60	1.34	1.26	1.33	1.53
6	3713	64	22.1	13.3	8.79	6.39	5.17	4.47	4.47
7	8961	128	68.1	40.7	24.8	16.2	11.9	8.89	7.56
8	21249	256	201	119	66.1	40.1	28.0	18.6	13.5
9	49665	512	575	379	169	106	71.6		28.0
10	114689	1024	1630			275	179		62.6

Table 6.22. *Convection-diffusion problem, uniformly refined sparse grids in three dimensions, timing of the iterative solution on Parnass2.*

level	time nodes	1/h	processors							
			1	4	16	32	64	128	256	512
2	81	4	0.04	0.06	0.10	0.14	0.16			
3	225	8	0.25	0.20	0.38	0.54	0.57	0.59		
4	593	16	1.31	0.71	0.89	1.24	1.53	1.87	2.34	3.26
5	1505	32	10.0	3.66	3.03	3.67	5.75	5.28	7.17	9.97
6	3713	64	52.5	20.1	12.8	11.9	13.1	15.4	20.5	29.9
7	8961	128	158	58.2	29.6	22.6	20.9	23.0	27.9	40.6
8	21249	256	473	192	66.6	44.9	67.6	32.2	36.7	94.0
9	49665	512	1426	396	156	95.4	62.0	48.6	48.2	102.2
10	114689	1024	4112				125	85.9	68.5	

Table 6.23. *Convection-diffusion problem, uniformly refined sparse grids in three dimensions, timing of the iterative solution on a Cray T3E-600.*

of the Alpha processor and even more the Power PC at 300 MHz respectively at 332 MHz, but the timings seem to be a combination of compiler technology, memory performance and the integer arithmetic fraction of the code.

If we fix the amount of work, that is the number of nodes per processor, we obtain the scaleup. Comparing a time at one level l and a number of processors p with the time of one level finer $l + 1$ and $2p$ processors, we obtain very good scaling of the method, as long as we do not consider the largest numbers of processors. The limits in the processor numbers are in this case not only a question of network scalability, but also an algorithmical issue of sparse grid

level	time nodes	1/h	processors 1	4	4 uni	16	64	256	512
4	593	16	0.67	0.47	0.42	0.46	0.54	0.80	0.66
5	1505	32	3.97	2.15	2.01	1.43	1.57	8.56	2.64
6	3713	64	23.33	11.14	10.81	5.70	4.80	19.63	9.34
7	8961	128	71.92	34.06	32.20	11.60	7.47	9.00	11.60
8	21249	256	208.8	90.62	86.40	27.33	12.41	10.89	13.75
9	49665	512	619.0	245.8	233.6	70.94	27.76	24.84	47.10
10	114689	1024	1845	740.5	693.0	189.2	66.36	45.92	62.85
11	262145	2048			2184	568.2	149.2	99.44	86.30
12	593921	4096				1766	343.2	143.0	150.6
13	1335297	8192					888.1	275.0	222.3

Table 6.24. *Convection-diffusion problem, uniformly refined sparse grids in three dimensions, timing of the iterative solution on ASCI Blue Pacific.*

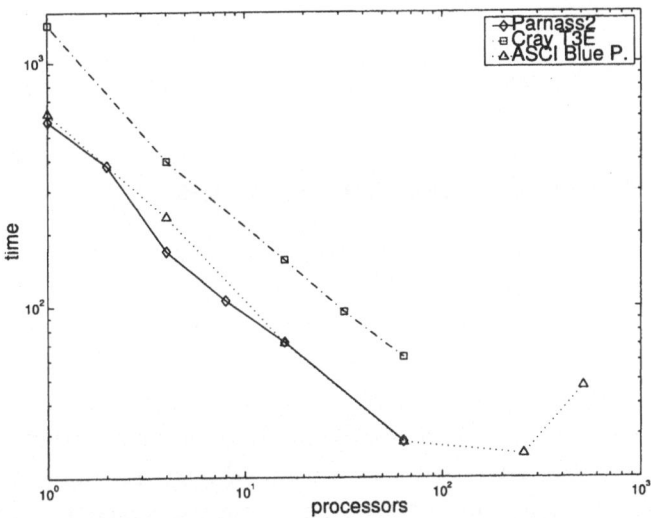

Figure 6.6. *Convection-diffusion problem, uniformly refined sparse grids in three dimensions, timing of the iterative solution on Parnass2, ASCI Blue Pacific and a Cray T3E-600.*

partitions. Note that in this case of uniform grid refinement, some a priori partition schemes would be superior to our dynamic partitioning scheme, and the communication free combination technique of discretisation would be even more efficient. However, our dynamic load balancing scheme performs well and

| time | | processors | | | | | |
nodes	1/h	1	2	4	8	16	32
81	4	0.03	0.04	0.05	0.07	0.11	
201	8	0.07	0.05	0.05	0.07	0.08	0.09
411	16	0.21	0.13	0.12	0.13	0.17	0.20
711	32	0.78	0.48	0.38	0.36	0.41	0.51
1143	64	2.60	1.49	1.06	0.93	0.92	1.14
1921	128	8.69	5.99	3.70	2.88	2.70	2.83
3299	256	39.3*	20.7	13.8	9.62	7.79	7.32
6041	512	177*	91.0	56.8	39.5	28.6	22.0
11787	1024	949*	525	271	177	138	88.2
22911	2048			1280	761	660	358

Table 6.25. *Convection-diffusion problem, adaptively refined sparse grids in three dimensions, timing of the iterative solution on Parnass2.*

introduces only a little overhead and results in good partitions, see Figure 6.5.

6.3.2 Adaptively Refined Sparse Grids

In this last experimental section we consider adaptively refined sparse grids for a problem with singularities, where the sparse grids are refined toward a singularity located in the lower left corner. Table 6.25 depicts times in the adaptive case. These numbers give the wall clock times for the solution of the equation system again, now on different levels of adaptive grids and on different numbers of processors. Due to the solution-dependent adaptive refinement criterion, the single processor version contained slightly more nodes, indicated by an asterisk *. For the same reason, the equation systems have been solved up to rounding error instead of the weaker discretisation error condition in the uniform sparse grid experiment.

We obtain a good scaling, both in the number of unknowns and in the number of processors, i.e. the times are proportional to the number of unknowns for a fixed number of processors and are indirect proportional to the number of processors. Increasing the number of processors speeds up the computation accordingly. The parallel efficiencies are somewhat smaller than for the uniform refinement case, due to the imbalance in the tree of nodes. This is also the case for other parallel adaptive methods. Hence this parallelisation approach does perform very well, even in the range of higher number of processors 16 and 32, where several other strategies are not competitive.

As we saw in this chapter, space-filling curve partitioning does not only perform well theoretically, but also in practical experiments. We saw a range of results by parallel multigrid methods for the Poisson equation and for the Lamé equation of linear elasticity, for uniformly refined meshes and for adaptively refined meshes on the Parnass2 cluster of different configurations and on Cray T3Es and ASCI Blue Pacific, both with up to 1024 processors. Furthermore, parallel sparse grid solvers for convection-diffusion and the Poisson equation in up to ten dimensions on uniform and on adaptively refined sparse grids were compared.

We obtained extremely good scaleups for most of the experiments and speedups, even on the 1024 processor configurations. However, we have to mention that the extremely large processor configurations merely serve as an experimental setup to show scalability in this high range of processors, since as a non-US citizen it is rarely possible to be able to access these computers at all, let alone to occupy them with many computations. This might explain some of the omissions of data and experiments, because the scheduled computing time was simply exceeded.

However, we can clearly state that space-filling curves are very well suited for the load balancing problem. The direct comparisons of load balancing to solution times showed this impressively. Furthermore, the numbers demonstrate that the partitions and the dual graph cut sizes allow for highly efficient parallelisation of the codes, even in the regime of large numbers of processors. Note that graph partitioning heuristics often do not perform all that well for large numbers of processors, which can also be seen in chapter 4.4. It remains to compare these graph based methods on large scale parallel computers, which has been done for space-filling curves in this chapter.

Concluding Remarks and Outlook

In this book we have discussed the discretisation of elliptic boundary value problems, adaptive mesh refinement, multilevel methods for the solution of the equation systems and the parallelisation of these methods. We have proposed and analysed algorithms for parallel load-balancing of multilevel methods on adaptively refined meshes by space-filling curves. Furthermore, we have demonstrated the efficiency of the approach on a number of large parallel computers.

First of all, we presented a unifying framework of subspace correction methods. Based on an abstract splitting of function spaces, iterative schemes like additive and multiplicative Schwarz iterations can be defined. Multigrid and multilevel methods have been characterised by a nested sequence of spaces, whereas spaces defined on a geometric partition of the domain lead to domain decomposition methods. Multilevel methods allow for an efficient solution of large equation systems obtained by the discretisation of PDEs in $\mathcal{O}(n)$ or $\mathcal{O}(n \log n)$ operations for n equations, which follows from estimates on the abstract subspace splittings. The main motivation for the development of domain decomposition was the need for efficient iterative solvers on parallel computers so that the computational domain is partitioned and mapped onto several processors. However, multigrid methods could also run efficiently in parallel. We concluded the chapter with a discussion of sparse grid discretisations and the solution of equation systems derived therefrom.

In chapter 3 we discussed adaptively refined meshes. Since a multigrid $\mathcal{O}(n)$ computational complexity for n degrees of freedom is optimal, there are only two ways for a substantial improvement of a PDE solver: parallelisation and adaptive mesh refinement. The chapter covered a variety of techniques of posteriori error estimation, adaptive mesh refinement, and representation of adaptively meshes with key based addressing and hash storage.

For the purpose of load balancing, space-filling curves were introduced in chapter 4. It was proven that the space-filling curve partitions were indeed well suited for parallel computing and the cut sizes of the meshes were of quasi-optimal size. The proof was generalised to adaptively refined meshes with criteria of β and γ-adaptive families of meshes. The results showed that usual adaptive refinement as well as best n-term approximation for solutions with point-singularities allow for quasi-optimal space-filling curve partitions. Furthermore, the refinement toward higher dimensional manifolds like edge singularities may also lead to quasi-optimal partitions, dependent on the dimension d of the domain and in the case of best n-term approximation also on some parameters of the underlying Besov spaces. In order to construct good partitions in the remaining cases, the construction of anisotropic space-filling curves was proposed. The chapter concluded with a discussion of sparse grid partitions.

In chapter 5 we combined multilevel methods, adaptive mesh refinement and parallelisation. Linear complexity multigrid methods on adaptively refined meshes, scalable parallel multigrid methods and finally parallel methods on adaptively refined meshes were discussed. For this purpose we proposed space-filling curve partitioning, key based addressing and a parallel hash storage scheme.

In the final chapter 6, numerical experiments showed how space-filling curve partitions can be used within an adaptive multilevel method and a sparse grid discretisation. A range of applications and parallel computer platforms demonstrated that the asymptotic estimates on graph cut sizes of space-filling curves were indeed useful, also for realistic test cases.

We can conclude that the concept of space-filling curves does indeed work well, leads to parallel and linear complexity partitioning algorithms, simplifies implementation details, and allows for a first time for real dynamic load balancing of adaptive computations.

There are still many open problems in the area of parallel numerical methods for partial differential equations with adaptive mesh refinement and efficient solvers. Since we covered only elliptic boundary value problems of second order, many generalisations are possible: time dependent parabolic problems with adaptive mesh refinement and de-refinement can be considered with space-filling curve partitioning and multilevel solvers at each time-step. Alternatively, discretisations both in space and time can be used where anisotropic space-filling curves would be more appropriate for load balancing. Hyperbolic problems can be treated in a similar way, see also Griebel and Zumbusch [153]. The linear solvers discussed so far can be used for the solution of PDEs with nonlinear operators or eigenvalue problems with outer or inner-loop linearisation. Of course, also other types of PDEs like Maxwell or Schrödinger equations can be solved in parallel with space-filling curve partitioning of adaptively refined meshes.

The mesh partitioning with space-filling curves was based on a strict load balancing, where processor i obtains nodes $\left[i\frac{n}{p}, (i+1)\frac{n}{p}\right)$ of the space-filling curve order. However, the cut ratios can be improved by a relaxed condition, as we saw for some counter examples in chapter 4.3. A best cut ratio can be chosen for a shift of the intervals on the closed space-filling curve or by taking into account a load imbalance. Furthermore, the partitions can be improved by some local optimisation methods like a few steps of the Kernighan Lin method, see also Ellerbrake [114]. The space-filling curve partitioning can also be applied to unstructured meshes, as in chapter 4, to patch-wise refined meshes and to further hybrid mesh constructions.

Bibliography

[1] V. I. AGOSHKOV AND V. I. LEBEDEV, *Poincaré-Steklov operators and domain decomposition methods in variational problems*, Vychisl. Protsessy Sist., 2 (1985), pp. 173–227.

[2] M. AINSWORTH AND J. T. ODEN, *A Posteriori Error Estimation in Finite Element Analysis*, J. Wiley, New York, 2000.

[3] J. ALBER AND R. NIEDERMEIER, *On multidimensional curves with Hilbert property*, Theory Comput. Systems, 33 (2000), pp. 295–312.

[4] P. R. AMESTOY, I. S. DUFF, AND J.-Y. L'EXCELLENT, *Multifrontal parallel distributed symmetric and unsymmetric solvers*, Comput. Methods Appl. Mech. Engrg., 184 (2000), pp. 501–520.

[5] E. ANDERSON, Z. BAI, C. BISCHOF, J. DEMMEL, J. DONGARRA, J. D. CROZ, A. GREENBAUM, S. HAMMARLING, A. MCKENNEY, S. OSTROUCHOV, AND D. SORENSEN, *LAPACK Users' Guide*, SIAM, Philadelphia, 1992.

[6] T. ASANO, D. RANJAN, T. ROOS, E. WELZL, AND P. WIDMAYER, *Space-filling curves and their use in the design of geometric data structures*, Theor. Comput. Sci., 181 (1997), pp. 3–15.

[7] C. ASHCRAFT, S. EISENSTAT, AND J. W.-H. LIU, *A fan-in algorithm for distributed sparse numerical factorization*, SIAM J. Sci. Statist. Comput., 11 (1990), pp. 593–599.

[8] C. ASHCRAFT AND R. GRIMES, *SPOOLES: An object-oriented sparse matrix library*, in Proc. of 9th SIAM Conf. on Parallel Processing for Scientific Computing (PP 99), B. Hendrickson et al., eds., San Antonio, Tx., 1999, SIAM.

[9] G. P. ASTRAKHANTSEV, *An iterative method for solving elliptic net problems*, Zh. Vychisl. Mat. Mat. Fiz., 11 (1971), pp. 439–448.

[10] O. AXELSSON AND P. S. VASSILEVSKI, *Algebraic multilevel preconditioning methods. I.*, Numer. Math., 56 (1989), pp. 157–177.

[11] K. I. BABENKO, *Approximation of periodic functions of many variables by trigonometric polynomials*, Soviet Math. Dokl., 1 (1960), pp. 513–516.

[12] I. BABUŠKA, *Über Schwarzsche Algorithmen in partiellen Differentialgleichungen der mathematischen Physik*, Z. Angew. Math. Mech., 37 (1957), pp. 243–245.

[13] I. BABUŠKA AND A. MILLER, *A feedback finite element method with a posteriori error estimation: I. The finite element method and some basic properties of the a posteriori error estimator*, Comput. Methods Appl. Mech. Engrg., 61 (1987), pp. 1–40.

[14] I. BABUŠKA AND W. C. RHEINBOLDT, *Error estimates for adaptive finite element computations*, SIAM J. Numer. Anal., 15 (1978), pp. 736–754.

[15] I. BABUŠKA, T. STROUBOULIS, D. DATTA, AND S. GANGARAJ, *What do we want*

and what do we have in a posteriori estimates in the FEM, in The mathematics of finite elements and applications X, MAFELAP 1999. Proc. of the 10th conf., Brunel Univ., Uxbridge, J. R. Whiteman, ed., Elsevier, Amsterdam, 2000, pp. 163–180.

[16] N. S. BACHVALOV, *On the convergence of a relaxation method with natural constraints on the elliptic operator*, Zh. Vychisl. Mat. Mat. Fiz., 6 (1966), pp. 861–883.

[17] R. BALDER, *Adaptive Verfahren für elliptische und parabolische Differentialgleichungen auf dünnen Gittern*, Doktorarbeit, TU München, Inst. für Informatik, 1994.

[18] R. E. BANK, *PLTMG: A Software Package for Solving Elliptic Partial Differential Equations, Users' Guide 8.0*, Frontiers in Applied Mathematics, SIAM, Philadelphia, 1998.

[19] R. E. BANK AND T. F. DUPONT, *An optimal order process for solving elliptic finite element equations*, Math. Comp., 36 (1981), pp. 35–51.

[20] R. E. BANK, T. F. DUPONT, AND H. YSERENTANT, *The hierarchical basis multigrid method*, Numer. Math., 52 (1988), pp. 427–458.

[21] R. E. BANK, A. H. SHERMAN, AND A. WEISER, *Refinement algorithms and data structures for regular local mesh refinement*, in Scientific Computing, R. Stepleman, ed., IMACS, North-Holland, Amsterdam, 1983, pp. 3–17.

[22] R. E. BANK AND A. WEISER, *Some a posteriori error estimators for elliptic partial differential equations*, Math. Comp., 44 (1985), pp. 283–301.

[23] R. E. BANK AND J. XU, *The hierarchical basis multigrid method and incomplete LU decompostion*, in Domain decomposition methods 7, vol. 180 of Contemp. Math., AMS, Providence, Rhode Island, 1994, pp. 163–173.

[24] S. T. BARNARD AND H. D. SIMON, *A fast multilevel implementation of recursive spectral bisection for partitioning unstructured problems*, Concurrency: Practice and Experience, 6 (1994), pp. 101–107.

[25] J. J. BARTHOLDI AND L. K. PLATZMAN, *Heuristics based on spacefilling curves for combinatorial optimization problems in Euclidean space.*, Manage. Sci., 34 (1988), pp. 291–305.

[26] P. BASTIAN, *Parallele Adaptive Mehrgitterverfahren*, Skripten zur Numerik, Teubner, Stuttgart, 1996.

[27] ——, *Load balancing for adaptive multigrid methods*, SIAM J. Sci. Comput., 19 (1998), pp. 1303–1321.

[28] P. BASTIAN, J. H. FERZIGER, G. HORTON, AND J. VOLKERT, *Adaptive multigrid solution of the convection-diffusion equation on the DIRMU processor*, in Robust Multi-Grid Methods, Proc. 4th GAMM-Semin., Kiel 1988, W. Hackbusch, ed., vol. 23 of Notes on Numerical Fluid Mechanics, Vieweg, Braunschweig, 1989, pp. 27–36.

[29] P. BASTIAN, S. LANG, AND K. ECKSTEIN, *Parallel adaptive multigrid methods in plane linear elasticity problems*, Numer. Linear Algebra Appl., 4 (1997), pp. 153–176.

[30] R. BECKER, C. JOHNSON, AND R. RANNACHER, *Adaptive error control for multigrid finite element methods*, Computing, 55 (1995), pp. 271–288.

[31] R. BECKER AND R. RANNACHER, *An optimal control approach to a posteriori error estimation in finite element methods*, in Acta Numerica, A. Iserles, ed., vol. 10, Cambridge Univ. Press, Cambridge, 2001, pp. 1–102.

[32] I. BEICHL AND F. SULLIVAN, *Interleave in peace, or interleave in pieces*, IEEE Computational Science & Engineering, 5 (1998), pp. 92–96.

[33] M. J. BERGER AND S. BOKHARI, *A partitioning strategy for nonuniform problems*

on multiprocessors, IEEE Trans. Comput., C-36 (1987), pp. 570–580.

[34] M. J. BERGER AND P. COLELLA, *Local adaptive mesh refinement for shock hydrody-namics*, J. Comput. Phys., 82 (1989), pp. 64–84.

[35] M. J. BERGER AND J. OLIGER, *Adaptive mesh refinement for hyperbolic partial differential equations*, J. Comput. Phys., 53 (1984), pp. 484–512.

[36] M. J. BERGER AND I. RIGOUTSOS, *An algorithm for point clustering and grid generation*, IEEE Trans. Systems Man Cybernet., 21 (1991), pp. 1278–1286.

[37] C. BERNADI, Y. MADAY, AND A. PATERA, *A new nonconforming approach to domain decomposition: The mortar method*, in Nonlinear Partial Differential Equations and their Applications, H. Brezis and J. L. Lions, eds., Pitman, London, 1989.

[38] D. BERTSIMAS AND M. GRIGNI, *On the spacefilling curve heuristic for the Euclidean traveling salesman problem*, tech. report, Massachusetts Institute of Technology, Cambridge, MA, 1988.

[39] J. BEY, *Analyse und Simulation eines Konjugierte-Gradienten-Verfahrens mit einem Multilevel Präkonditionierer zur Lösung dreidimensionaler elliptischer Randwertprobleme für massiv parallele Rechner*, Diplomarbeit, RWTH Aachen, 1991.

[40] ———, *Tetrahedral grid refinement*, Computing, 55 (1995), pp. 355–378.

[41] ———, *Simplicial grid refinement: On Freudenthal's algorithm and the optimal number of congruence classes*, Numer. Math., 85 (2000).

[42] K. BIRKEN, *Ein Parallelisierungskonzept für adaptive, numerische Berechnungen*, Diplomarbeit, Universität Erlangen-Nürnberg, 1993.

[43] K. BIRKEN AND C. HELF, *A dynamic data model for parallel adaptive PDE solvers*, in Proc. of HPCN Europe 1995, B. Hertzberger and G. Serazzi, eds., vol. 919 of Lecture Notes in Computer Science, Milan, Italy, 1995, Springer, Berlin, Heidelberg.

[44] P. E. BJØRSTAD AND J. MANDEL, *On the spectra of sums of orthogonal projections with applications to parallel computing*, BIT, 31 (1991), pp. 76–88.

[45] P. E. BJØRSTAD AND O. B. WIDLUND, *Iterative methods for the solution of elliptic problems on regions partitioned into substructures*, SIAM J. Numer. Anal., 23 (1986), pp. 1097–1120.

[46] S. H. BOKHARI, T. W. CROCKETT, AND D. N. NICOL, *Parametric binary dissection*, Tech. Report 93-39, ICASE, 1993.

[47] C. BÖRGERS, *The Neumann-Dirichlet domain decomposition method with inexact solvers on the subdomains*, Numer. Math., 55 (1989), pp. 123–136.

[48] F. A. BORNEMANN, *An adaptive multilevel approach to parabolic equations III: 2D error estimation and multilevel preconditioning.*, IMPACT Comput. Sci. Engrg., 4 (1992), pp. 1–45.

[49] ———, *Interpolation spaces and optimal multilevel preconditioners*, in Domain decomposition methods 7, vol. 180 of Contemp. Math., AMS, Providence, Rhode Island, 1994, pp. 3–8.

[50] F. A. BORNEMANN, B. ERDMANN, AND R. KORNHUBER, *A posteriori error estimates for elliptic problems in two and three space dimensions*, SIAM J. Numer. Anal., 33 (1996), pp. 1188–1204.

[51] J.-F. BOURGAT, R. GLOWINSKI, P. L. TALLEC, AND M. VIDRASCU, *Variational formulation and algorithm for trace operator in domain decomposition calculations*, in Domain decomposition methods 2, T. F. Chan, R. Glowinski, J. Périaux, and O. B. Widlund, eds., SIAM, Philadelphia, 1989, pp. 3–16.

[52] D. BRAESS, *The contraction number of a multigrid method for solving the Poisson equation*, Numer. Math., 37 (1981), pp. 387–404.

[53] ——, *Towards algebraic multigrid for elliptic problems of second order*, Computing, 55 (1995), pp. 379–393.

[54] ——, *Finite Elements*, Cambridge Univ. Press, Cambridge, 1997.

[55] D. BRAESS AND C. BLÖMER, *A multigrid method for a parameter dependent problem in solid mechanics*, Numer. Math., 57 (1990), pp. 747–761.

[56] H. BRAKHAGE, *Über die numerische Behandlung von Intregralgleichungen nach der Quadraturformelmethode*, Numer. Math., 2 (1960), pp. 183–196.

[57] J. H. BRAMBLE, *Multigrid Methods*, vol. 294 of Pitman Research Notes in Mathematical Sciences, Longman Scientific & Technical, Essex, 1993.

[58] J. H. BRAMBLE AND J. E. PASCIAK, *New estimates for multilevel algorithms including the V-cycle*, Math. Comp., 60 (1993), pp. 447–471.

[59] J. H. BRAMBLE, J. E. PASCIAK, AND A. H. SCHATZ, *An iterative method for elliptic problems on regions partitioned into substructures*, Math. Comp., 46 (1986), pp. 361–369.

[60] ——, *The construction of preconditioners for elliptic problems by substructuring. IV.*, Math. Comp., 53 (1989), pp. 1–24.

[61] J. H. BRAMBLE, J. E. PASCIAK, J. WANG, AND J. XU, *Convergence estimates for multigrid algorithms without regularity assumptions*, Math. Comp., 57 (1991), pp. 23–45.

[62] ——, *Convergence estimates for product iterative methods with applications to domain decomposition*, Math. Comp., 57 (1991), pp. 1–21.

[63] J. H. BRAMBLE, J. E. PASCIAK, AND J. XU, *Parallel multilevel preconditioners*, Math. Comp., 55 (1990), pp. 1–22.

[64] A. BRANDT, *Multi-level adaptive technique (MLAT) for fast numerical solution to boundary value problems*, in Proc. of the 3rd Int. Conf. on Numerical Methods in Fluid Mechanics, Univ. Paris 1972, H. Cabannes and R. Teman, eds., vol. 18 of Lecture Notes in Physics, Springer, Berlin, Heidelberg, 1973, pp. 82–89.

[65] ——, *Multi-level adaptive solutions to boundary-value problems*, Math. Comp., 31 (1977), pp. 333–390.

[66] ——, *Multigrid solvers on parallel computers*, in Elliptic Problem Solvers, Proc. Conf., Santa Fe, M. H. Schultz, ed., Academic Press, San Diego, CA, 1981, pp. 39–83.

[67] ——, *Guide to multigrid development*, in Multigrid Methods, Proc. Conf., Köln-Porz, W. Hackbusch and U. Trottenberg, eds., vol. 960 of Lecture Notes in Mathematics, Springer, Berlin, Heidelberg, 1982, pp. 220–312.

[68] ——, *Algebraic multigrid theory: The symmetric case*, Appl. Math. Comput., 19 (1986), pp. 23–56.

[69] A. BRANDT AND B. DISKIN, *Multigrid solvers on decomposed domains*, in Domain decomposition methods 6, vol. 157 of Contemp. Math., AMS, Providence, Rhode Island, 1994, pp. 135–155.

[70] S. C. BRENNER, *A nonconforming mixed multigrid method for the pure displacement problem in planar linear elasticity*, SIAM J. Numer. Anal., 30 (1993), pp. 116–135.

[71] ——, *Convergence of the multigrid V-cycle algorithms for second order boundary value problems without full elliptic regularity*, Math. Comp., 71 (2002), pp. 507–525.

[72] S. C. BRENNER AND L. R. SCOTT, *The Mathematical Theory of Finite Element*

Methods, Texts in Applied Mathematics, Springer, New York, 1994.

[73] W. L. BRIGGS, W. E. HENSON, AND S. F. McCORMICK, *A Multigrid Tutorial*, SIAM, Philadelphia, 2000.

[74] A. M. BRUASET, X. CAI, H. P. LANGTANGEN, G. T. LINES, K. SAMUELSSON, W. SHEN, A. TVEITO, AND G. ZUMBUSCH, *CPU-measurements of some numerical PDE-applications*, tech. report, Sintef Applied Mathematics, Oslo, Norway, 1997.

[75] E. BUGNION, T. ROOS, R. WATTENHOFER, AND P. WIDMAYER, *Space filling curves versus random walks*, in Proc. Algorithmic Foundations of Geographic Information Systems, vol. 1340 of Lecture Notes in Computer Science, Springer, Berlin, Heidelberg, 1997.

[76] T. N. BUI AND C. JONES, *Finding good approximate vertex and edge partitions is NP-hard*, Inf. Process. Lett., 42 (1992), pp. 153–159.

[77] O. BUNEMAN, *A compact non-iterative Poisson solver*, SUIPR report 294, Stanford, 1969.

[78] H.-J. BUNGARTZ, *Dünne Gitter und deren Anwendung bei der adaptiven Lösung der dreidimensionalen Poisson-Gleichung*, Doktorarbeit, TU München, Inst. für Informatik, 1992.

[79] ———, *Finite Elements of Higher order on Sparse Grids*, Habilitation, TU München, Inst. für Informatik, 1998.

[80] H.-J. BUNGARTZ AND T. DORNSEIFER, *Sparse grids: Recent developments for elliptic partial differential equations*, in Multigrid Methods V. Proc. of the 5th European multigrid conf., Stuttgar, W. Hackbusch and G. Wittum, eds., vol. 3 of Lecture Notes in Computational Science and Engineering, Springer, Berlin, Heidelberg, 1998, pp. 45–70.

[81] H.-J. BUNGARTZ, T. DORNSEIFER, AND C. ZENGER, *Tensor product approximation spaces for the efficient numerical solution of partial differential equations*, in Proc. Int. Workshop on Scientific Computations, Nova Science Publishers, Konya, 1999. to appear.

[82] H.-J. BUNGARTZ, M. GRIEBEL, D. RÖSCHKE, AND C. ZENGER, *Two proofs of convergence for the combination technique for the efficient solution of sparse grid problems*, in Domain decomposition methods 7, D. E. Keyes and J. Xu, eds., vol. 180 of Contemp. Math., AMS, Providence, Rhode Island, 1994, pp. 15–20.

[83] F. CAO, J. R. GILBERT, AND S.-H. TENG, *Partitioning meshes with lines and planes*, Tech. Report CSL-96-01, Xerox Palo Alto Research Center, 1996.

[84] J. A. CAVENDISH, W. J. GORDON, AND C. A. HALL, *Ritz-Galerkin approximations in blending function spaces*, Numer. Math., 26 (1976), pp. 155–178.

[85] T. F. CHAN AND T. P. MATHEW, *Domain decomposition algorithms*, in Acta Numerica, A. Iserles, ed., vol. 3, Cambridge Univ. Press, Cambridge, 1994, pp. 61–143.

[86] G. CHOCHIA AND M. COLE, *Recursive 3D mesh indexing with improved locality*, in Proc. HPCN '97, no. 1225 in Lecture Notes in Computer Science, Springer, Berlin, Heidelberg, 1997, pp. 1014–1015.

[87] G. CHOCHIA, M. COLE, AND T. HEYWOOD, *Implementing the hierarchical PRAM on the 2D mesh: Analyses and experiments*, in Proc. 7th IEEE Symposium on Parallel and Distributed Processing (SPDP '95), IEEE, New York, 1995.

[88] P. G. CIARLET, *The Finite Element Method for Elliptic Problems*, North-Holland, Amsterdam, 1978.

[89] ——, *Three-Dimensional Elasticity*, vol. 1 of Mathematical Elasticity, North-Holland, Amsterdam, 1988.

[90] W. CLARKE, *Key-based parallel adaptive refinement for FEM*, bachelor thesis, Australian National Univ., Dept. of Engineering, 1996.

[91] A. COHEN, W. DAHMEN, AND R. DEVORE, *Adaptive wavelet methods for elliptic operator equations: Convergence rates*, Math. Comp., 70 (2001), pp. 27–75.

[92] *Proc. of the Copper Mountain Multigrid Conf. 1–11*, Appl. Math. Comput./ Marcel Dekker/ SIAM/ NASA (Langley)/ Electronic Trans. Numer. Anal./ Numer. Linear Algebra Appl., 1981/ 1986/ 1987–2003 (biannually).

[93] A. W. CRAIG, J. Z. ZHU, AND O. C. ZIENKIEWICZ, *A-posteriori error estimation, adaptive mesh refinement and multigrid methods using hierarchical finite element bases*, in The mathematics of finite elements and applications V, MAFELAP 1984, Proc. 5th Conf., Uxbridge, 1985, pp. 587–594.

[94] G. CYBENKO, *Load balancing for distributed memory multiprocessors*, J. Parallel Distributed Comput., 7 (1989), pp. 279–301.

[95] S. DAHLKE, W. DAHMEN, AND R. DEVORE, *Nonlinear approximation and adaptive techniques for solving elliptic operator equations*, in Multiscale Techniques for PDEs, W. Dahmen, A. Kurdilla, and P. Oswald, eds., Academic Press, San Diego, CA, 1997, pp. 237–284.

[96] S. DAHLKE, W. DAHMEN, R. HOCHMUTH, AND R. SCHNEIDER, *Stable multiscale bases and local error estimation for elliptic problems*, Appl. Numer. Math., 1 (1997), pp. 21–47.

[97] W. DAHMEN, *Wavelet and multiscale methods for operator equations*, in Acta Numerica, A. Iserles, ed., vol. 6, Cambridge Univ. Press, Cambridge, 1997, pp. 55–228.

[98] W. DAHMEN AND A. KUNOTH, *Multilevel preconditioning*, Numer. Math., 63 (1992), pp. 315–344.

[99] M. DAUGE, *Elliptic Boundary Value Problems on Corner Domains*, vol. 1341 of Lecture Notes in Mathematics, Springer, Berlin, Heidelberg, 1988.

[100] J. DE KEYSER AND D. ROOSE, *Partitioning and mapping adaptive multigrid hierarchies on distributed memory computers*, Tech. Report TW 166, Univ. Leuven, Dept. Computer Science, 1992.

[101] R. DEVORE, *Nonlinear approximation*, in Acta Numerica, A. Iserles, ed., vol. 7, Cambridge Univ. Press, Cambridge, 1998, pp. 51–150.

[102] R. DEVORE AND V. A. POPOV, *Interpolation spaces and nonlinear approximation*, in Function Spaces and Applications, Proc. US-Swed. Semin., Lund, M. Cwikel, J. Peetre, Y. Sagher, and H. Wallin, eds., vol. 1302 of Lecture Notes in Mathematics, Springer, Berlin, Heidelberg, 1988, pp. 191–205.

[103] F. DOBRIAN, G. KUMFERT, AND A. POTHEN, *The design of sparse direct solvers using object-oriented techniques*, in Advances in Software Tools in Scientific Computing, H. P. Langtangen, A. M. Bruaset, and E. Quak, eds., no. 10 in Lecture Notes in Computational Science and Engineering, Springer, Berlin, Heidelberg, 2000, ch. 3, pp. 89–131.

[104] *Proc. of Domain Decomposition Methods in Science and Engineering 1–15*, SIAM/ AMS/ J. Wiley/ DDM.org, 1988–2003.

[105] W. DÖRFLER, *A convergent adaptive algorithm for Poisson's equation*, SIAM J. Numer. Anal., 33 (1996), pp. 1106–1124.

[106] C. C. DOUGLAS, *Parallel multilevel and multigrid methods*, SIAM News, 25 (1992), pp. 14–15.

[107] ——, *Caching in with multigrid algorithms: problems in two dimensions*, Paral. Alg. Appl., 9 (1996), pp. 195–204.

[108] M. DRYJA, *An algorithm with a capacitance matrix for a variational-difference Dirichlet problem*, in Variational-Difference Methods in Mathematical Physics, IV. All-Union Conf., Novosibirsk, G. I. Marchuk, ed., USSR Academy of Sciences, Novosibirsk, 1981, pp. 63–73.

[109] M. DRYJA, B. F. SMITH, AND O. B. WIDLUND, *Schwarz analysis of iterative substructuring algorithms for elliptic problems in three dimensions*, SIAM J. Numer. Anal., 31 (1994), pp. 1662–1694.

[110] M. DRYJA AND O. B. WIDLUND, *Towards a unified theory of domain decomposition algorithms for elliptic problems*, in Domain decomposition methods 3, T. F. Chan, R. Glowinski, J. Périaux, and O. B. Widlund, eds., SIAM, Philadelphia, 1990, pp. 3–21.

[111] ——, *Additive Schwarz methods for elliptic finite element problems in three dimensions*, in Domain decomposition methods 5, T. F. Chan, D. E. Keyes, G. Meurant, J. S. Scroggs, and R. G. Voigt, eds., SIAM, Philadelphia, 1992, pp. 3–18.

[112] ——, *Schwarz methods of Neumann-Neumann type for three-dimensional elliptic finite element problems*, Commun. Pure Appl. Math., 48 (1995), pp. 121–155.

[113] J. DUBINSKI, *A parallel tree code*, New Astronomy, 1 (1996), pp. 133–147.

[114] M. ELLERBRAKE, *Eine schnelle näherungsweise Lösung des Euklidischen Travelling Salesman Problems mittels raumfüllender Kurven*, Diplomarbeit, Inst. für Angew. Math., Universität Bonn, 2000.

[115] H. ESSER AND K. NIEDERDRENK, *Nichtäquidistante Diskretisierungen von Randwertaufgaben*, Numer. Math., 35 (1980), pp. 465–478.

[116] C. FARHAT AND F.-X. ROUX, *Implicit parallel processing in structural mechanics*, Comput. Mech. Adv., 2 (1994).

[117] R. P. FEDORENKO, *A relaxation method for solving elliptic difference equations*, U.S.S.R. Comput. Math. Math. Phys., 1 (1962), pp. 1092–1096.

[118] ——, *The speed of convergence of one iterative process*, U.S.S.R. Comput. Math. Math. Phys., 4 (1964), pp. 227–235.

[119] M. FRIGO AND S. G. JOHNSON, *FFTW: An adaptive software architecture for the FFT*, in Proc. IEEE Intl. Conf. on Acoustics, Speech, and Signal Processing, vol. 3, 1998, pp. 1381–1384.

[120] C. FU, X. JIAO, AND T. TANG, *Efficient sparse LU factorization with partial pivoting on distributed memory architectures*, IEEE Trans. Parallel Distributed Systems, 9 (1998), pp. 109–125.

[121] J. FUHRMANN, *On the convergence of algebraically defined multigrid methods*, tech. report, Inst. für Angew. Analysis und Stochastik, Berlin, 1992.

[122] K. A. GALLIVAN, B. A. MARSOLF, AND H. A. G. WIJSHOFF, *Solving large nonsymmetric sparse linear systems using MCSPARSE*, Parallel Comput., 22 (1996), pp. 1291–1333.

[123] J. GAO AND J. M. STEELE, *General spacefilling curve heuristics and limit theory for the traveling salesman problem*, J. Complexity, 10 (1994), pp. 230–245.

[124] A. M. GARSIA, *Combinatorial inequalities and smoothness of functions*, Bull. Am.

Math. Soc., 82 (1976), pp. 157–170.

[125] A. GEORGE AND J. W. H. LIU, *Computer Solution of Large, Sparse, Positive Definite Systems*, Prentice Hall, New Jersey, 1981.

[126] ——, *The evolution of the minimum degree ordering algorithm*, SIAM Rev., 31 (1989), pp. 1–19.

[127] B. GHOSH, S. MUTHUKRISHNAN, AND M. H. SCHULTZ, *First and second order diffusive methods for rapid, coarse, distributed load balancing*, in 8th Annual ACM Symposium on Parallel Algorithms and Architectures, SPAA, 1996, pp. 72–81.

[128] G. H. GOLUB AND C. F. V. LOAN, *Matrix Computations*, Johns Hopkins Univ. Press, Baltimore, MD, 3rd ed., 1996.

[129] G. H. GOLUB AND D. MAYERS, *The use of pre-conditioning over irregular regions*, in Computing Methods in Applied Sciences and Engineering VI, Proc. 6th Int. Symp., Versailles, R. Glowinski and J.-L. Lions, eds., North-Holland, Amsterdam, 1984, pp. 3–14.

[130] C. GOTSMAN AND M. LINDENBAUM, *On the metric properties of discrete space-filing curves*, IEEE Trans. Image Processing, 5 (1996), pp. 794–797.

[131] T. GRAUSCHOPF, M. GRIEBEL, AND H. REGLER, *Additive multilevel preconditioners based on bilinear interpolation, matrix-dependent geometric coarsening and algebraic multigrid coarsening for second-order elliptic PDEs*, Appl. Numer. Math., 23 (1997), pp. 63–96.

[132] B. GREER AND G. HENRY, *From micro-ops to teraflops*, in Proc. Supercomputing 97, 1997.

[133] M. GRIEBEL, *Parallel multigrid methods on sparse grids*, in Multigrid Methods III, Proc. 3rd Eur. Conf., Bonn, W. Hackbusch and U. Trottenberg, eds., vol. 98 of ISNM, Birkhäuser, Basel, 1991, pp. 211–221.

[134] ——, *A parallelizable and vectorizable multi-level algorithm on sparse grids*, in Parallel Algorithms for partial differential equations, Proc. 6th GAMM-Semin., Kiel, W. Hackbusch, ed., vol. 31 of Notes on Numerical Fluid Mechanics, Vieweg, Braunschweig, 1991, pp. 94–100.

[135] ——, *The combination technique for the sparse grid solution of PDEs on multiprocessor machines*, Parallel Processing Letters, 2 (1992), pp. 61–70.

[136] ——, *Eine Kombinationstechnik für die Lösung von Dünn-Gitter-Problemen auf Multiprozessor-Maschinen*, in Numerische Algorithmen auf Transputer-Systemen, G. Bader, R. Rannacher, and G. Wittum, eds., Skripten zur Numerik, Teubner, Stuttgart, 1993, pp. 67–79.

[137] ——, *Multilevel algorithms considered as iterative methods on semidefinite systems*, SIAM J. Sci. Statist. Comput., 15 (1994), pp. 547–565.

[138] ——, *Multilevelmethoden als Iterationsverfahren über Erzeugendensystemen*, Skripten zur Numerik, Teubner, Stuttgart, 1994.

[139] ——, *Parallel point-oriented multilevel methods*, in Multigrid methods IV. Proc. of the 4th European multigrid conf., Amsterdam, P. W. Hemker and P. Wesseling, eds., ISNM, Birkhäuser, Basel, 1994, pp. 215–232.

[140] ——, *Adaptive sparse grid multilevel methods for elliptic PDEs based on finite differences*, Computing, 61 (1998), pp. 151–179.

[141] M. GRIEBEL, S. KNAPEK, G. ZUMBUSCH, AND A. CAGLAR, *Numerische Simulation in der Molekulardynamik. Numerik, Algorithmen, Parallelisierung, Anwendun-*

gen, Springer, Berlin, Heidelberg, 2004. to appear.

[142] M. GRIEBEL AND F. KOSTER, *Adaptive wavelet solvers for the unsteady incompressible Navier-Stokes equations*, in Advances in Mathematical Fluid Mechanics. Lecture Notes of the 6th Int. School "Mathematical Theory in Fluid Mechanics", Paseky, Czech Republic 1999, J. Malek, J. Necas, and M. Rokyta, eds., Springer, Berlin, Heidelberg, 2000, pp. 67–118.

[143] M. GRIEBEL AND T. NEUNHOEFFER, *Parallel point- and domain-oriented multilevel methods for elliptic PDE's on workstation networks*, J. Comput. Appl. Math., 66 (1996), pp. 267–278.

[144] M. GRIEBEL AND P. OSWALD, *Remarks on the abstract theory of additive and multiplicative Schwarz methods*, Tech. Report TUM-I9314, TU München, Inst. für Informatik, 1993.

[145] ——, *On additive Schwarz preconditioners for sparse grid discretizations*, Numer. Math., 66 (1994), pp. 449–463.

[146] ——, *On the abstract theory of additive and multiplicative Schwarz algorithms*, Numer. Math., 70 (1995), pp. 163–180.

[147] ——, *Tensor product type subspace splittings and multilevel iterative methods for anisotropic problems*, Adv. Comput. Math., 4 (1995), pp. 171–206.

[148] M. GRIEBEL, P. OSWALD, AND T. SCHIEKOFER, *Sparse grids for boundary integral equations*, Numer. Math., 83 (1999), pp. 279–312.

[149] M. GRIEBEL, M. SCHNEIDER, AND C. ZENGER, *A combination technique for the solution of sparse grid problems*, in Iterative Methods in Linear Algebra. Proc. of the IMACS int. symposium, Brussels, R. Beauwens and P. D. Groen, eds., North-Holland, Amsterdam, 1992, pp. 263–281.

[150] M. GRIEBEL AND G. ZUMBUSCH, *Parnass: Porting gigabit-LAN components to a workstation cluster*, in 1. Workshop Cluster Computing 1997, W. Rehm, ed., no. CSR-97-05 in Chemnitzer Informatik Berichte, TU Chemnitz, 1997. also as Tech. Rep. 19, SFB 256, Univ. Bonn.

[151] ——, *Hash-storage techniques for adaptive multilevel solvers and their domain decomposition parallelization*, in Domain decomposition methods 10, J. Mandel, C. Farhat, and X.-C. Cai, eds., vol. 218 of Contemp. Math., AMS, Providence, Rhode Island, 1998, pp. 271–278.

[152] ——, *Parallel multigrid in an adaptive PDE solver based on hashing*, in Parallel Computing: Fundamentals, Applications and New Directions, E. D'Hollander, G. R. Joubert, F. J. Peters, and U. Trottenberg, eds., no. 12 in Advances in Parallel Computing, Elsevier, Amsterdam, 1998, pp. 589–599.

[153] ——, *Adaptive sparse grids for hyperbolic conservation laws*, in Hyperbolic Problems: Theory, Numerics, Applications, M. Fey and R. Jeltsch, eds., vol. 1 of ISNM 129, Birkhäuser, Basel, 1999, pp. 411–422.

[154] ——, *Parallel adaptive subspace correction schemes with applications to elasticity*, Comput. Methods Appl. Mech. Engrg., 184 (2000), pp. 303–332.

[155] C. E. GROSCH, *Poisson solvers on large array computers*, in Proc. 1978 LANL Workshop on Vector and Parallel Processors, B. L. Buzbee and J. F. Morrison, eds., 1978.

[156] C. GROSSMANN AND H.-G. ROOS, *Numerik partieller Differentialgleichungen*, Studienbücher, Teubner, Stuttgart, 1994.

[157] S. GUATTERY AND G. L. MILLER, *On the quality of spectral separators*, SIAM J.

Matrix Anal. Appl., 19 (1998), pp. 701–719.

[158] I. GUSTAFSSON, *A class of first order factorization methods*, BIT, 18 (1978), pp. 142–156.

[159] G. HAASE, *Parallelisierung numerischer Algorithmen für partielle Differentialgleichungen*, Teubner, Stuttgart, Leipzig, 1999.

[160] G. HAASE, U. LANGER, AND A. MEYER, *A new approach to the Dirichlet domain decomposition method*, in Fifth Multigrid Seminar, Eberswalde 1990, S. Hengst, ed., Berlin, 1990, Karl-Weierstrass-Institut, pp. 1–59. Report R-MATH-09/90.

[161] ——, *Domain decomposition preconditioners with inexact subdomain solvers*, J. Numer. Lin. Alg. Appl., 1 (1991), pp. 27–42.

[162] G. HAASE, U. LANGER, A. MEYER, AND S. V. NEPOMNYASCHIKH, *Hierarchical extension operators and local multigrid methods in domain decomposition preconditioners*, East-West J. Numer. Math., 2 (1994), pp. 173–193.

[163] W. HACKBUSCH, *Ein iteratives Verfahren zur schnellen Auflösung elliptischer Randwertprobleme*, Tech. Report 76-12, Math. Inst., Universität zu Köln, 1976.

[164] ——, *A fast numerical method for elliptic boundary value problems with variable coefficients*, in 2nd GAMM-Conf. Numer. Methods Fluid Mech., Köln, E. H. Hirschel and W. Geller, eds., DFVLR, 1977, pp. 50–57.

[165] ——, *Multi-grid convergence theory*, in Multigrid Methods, Proc. Conf., Köln-Porz, W. Hackbusch and U. Trottenberg, eds., vol. 960 of Lecture Notes in Mathematics, Springer, Berlin, Heidelberg, 1982, pp. 177–219.

[166] ——, *Local defect correction method and domain decomposition techniques*, in Defect Correction Methods. Theory and Applications, K. Böhmer and H. J. Stetter, eds., Computing Suppl. 5, Springer, Vienna, 1984, pp. 89–113.

[167] ——, *Multi-Grid Methods and Applications*, Springer, Berlin, Heidelberg, 1985.

[168] ——, *Elliptic Differential Equations. Theory and Numerical Treatment*, Springer, New York, 1992.

[169] ——, *Iterative solution of large sparse systems of equations*, Springer, New York, 1994.

[170] W. HACKBUSCH AND S. A. SAUTER, *Composite finite elements for the approximation of PDEs on domains with complicated micro-structures*, Numer. Math., 75 (1997), pp. 447–472.

[171] G. HEBER, G. R. GAO, AND R. BISWAS, *Self-avoiding walks over adaptive unstructured grids*, Concurrency: Practice and Experience, 12 (2000), pp. 85–109.

[172] P. W. HEMKER, *Sparse-grid finite-volume multigrid for 3D-problems*, Adv. Comput. Math., 4 (1995), pp. 83–110.

[173] P. W. HEMKER AND B. KOREN, *A non-linear multigrid method for the steady Euler equations*, in Numerical Simulation of Compressible Euler Flows, A. Dervieux, B. van Leer, J. Périaux, and A. Rizzi, eds., vol. 26 of Notes on Numerical Fluid Mechanics, Vieweg, Braunschweig, 1989, pp. 175–196.

[174] B. HENDRICKSON AND R. LELAND, *An improved spectral graph partitioning algorithm for mapping parallel computations*, SIAM J. Sci. Statist. Comput., 16 (1995), pp. 452–469.

[175] D. HILBERT, *Über die stetige Abbildung einer Linie auf ein Flächenstück*, Mathematische Annalen, 38 (1891), pp. 459–460.

[176] R. HIPTMAIR, *Multilevel Preconditioning for Mixed Problems in Three Dimensions*,

Doktorarbeit, Math. Inst., Universität Augsburg, 1996.

[177] R. HOCHMUTH, S. KNAPEK, AND G. ZUMBUSCH, *Tensor products of Sobolev spaces and applications*, Tech. Report 685, Univ. Bonn, SFB 256, 2000.

[178] G. HORTON, *A multi-level diffusion method for dynamic load balancing*, Parallel Comput., 19 (1993), pp. 209–218.

[179] P. HOUSTON, R. RANNACHER, AND E. SÜLI, *A posteriori error analysis for stabilised finite element approximations of transport problems*, Comput. Methods Appl. Mech. Engrg., 190 (2000), pp. 1483–1508.

[180] J. HUNGERSHÖFER AND J.-M. WIERUM, *On the quality of partitions based on space-filling curves*, in Int. Conf. Computational Science ICCS, vol. 2331 of LNCS, Springer, Berlin, Heidelberg, 2002, pp. 36–45.

[181] M. T. JONES AND P. E. PLASSMANN, *Parallel algorithms for adaptive mesh refinement*, SIAM J. Sci. Comput., 18 (1997), pp. 686–708.

[182] G. E. KARNIADAKIS AND S. J. SHERWIN, *Spectral/hp Element Methods for CFD*, Oxford Univ. Press, New York, 1999.

[183] G. KARYPIS AND V. KUMAR, *Multilevel k-way graph partitioning for irregular graphs*, J. Parallel Distributed Comput., 48 (1998), pp. 96–129.

[184] D. W. KELLY, J. P. D. GAGO, O. C. ZIENKIEWICZ, AND I. BABUŠKA, *A posteriori error analysis and adaptive processes in the finite element method: Part I - error analysis*, Internat. J. Numer. Methods Engrg., 19 (1983).

[185] S. KNAPEK, *Approximation und Kompression mit Tensorprodukt-Multiskalenräumen*, Doktorarbeit, Universität Bonn, Inst. für Angew. Math., 2000.

[186] D. E. KNUTH, *The Art of Computer Programming*, vol. 3, Addison-Wesley, Reading, MA, 1975.

[187] S. R. KOHN AND S. B. BADEN, *A robust parallel programming model for dynamic non-uniform scientific computations*, Tech. Report CS94-354, UCSD, Dept. Computer Science, 1994.

[188] R. KORNHUBER AND R. ROITZSCH, *On adaptive grid refinement in the presence of internal or boundary layers*, IMPACT Comput. Sci. Engrg., 2 (1990), pp. 40–72.

[189] R. KORNHUBER AND H. YSERENTANT, *Multilevel methods for elliptic problems on domains not resolved by the coarse grid*, in Domain decomposition methods 7, vol. 180 of Contemp. Math., AMS, Providence, Rhode Island, 1994, pp. 49–60.

[190] F. KOSTER, *A proof of the consistency of the finite difference technique on sparse grids*, Computing, 65 (2001), pp. 247–261.

[191] ——, *Multiskalen-basierte Finite Differenzen Verfahren auf adaptiven dünnen Gittern*, Doktorarbeit, Universität Bonn, Inst. für Angew. Math., 2002.

[192] M.-H. LALLEMAND, H. STEVE, AND A. DERVIEUX, *Unstructured multigridding by volume agglomeration: Current status*, Comput. Fluids, 21 (1992), pp. 397–433.

[193] P. LE TALLEC, *Domain decomposition methods in computational mechanics*, Comput. Mech. Adv., 1 (1994), pp. 121–220.

[194] P. LE TALLEC, Y. H. DE ROECK, AND M. VIDRASCU, *Domain decomposition methods for large linearly elliptic three-dimensional problems*, J. Comput. Appl. Math., 34 (1991), pp. 93–117.

[195] H. LEBESGUE, *Sur les fonctions représentables analytiquement*, J. de Math., 6 (1905), pp. 139–216.

[196] C.-O. LEE, *Multigrid methods for the pure traction problem of linear elasticity: Mixed*

formulation, SIAM J. Numer. Anal., 35 (1998), pp. 121–145.

[197] P. LEINEN, *Ein schneller adaptiver Löser für elliptische Randwertprobleme auf Seriell-und Parallelrechnern*, Doktorarbeit, Universität Dortmund, 1990.

[198] M. LEMKE, *Multilevelverfahren mit selbstadaptiven Gitterverfeinerungen für Parallelrechner mit verteiltem Speicher*, Oldenbourg, Munich, 1994.

[199] M. LEMKE AND D. QUINLAN, *Fast adaptive composite grid methods on distributed parallel architectures*, Commun. Appl. Numer. Methods, 8 (1992), pp. 609–619.

[200] X. LI AND J. DEMMEL, *Making sparse Gaussian elimination scalable by static pivoting*, in Proc. of Supercomputing'98, 1998.

[201] C. B. LIEM, T. LU, AND T. M. SHIH, *The Splitting Extrapolation Method: A New Technique in Numerical Solution of Multidimensional Problems*, vol. 7 of Applied Mathematics, World Scientific, Singapore, 1995.

[202] P.-L. LIONS, *Interprétation stochastique de la méthode alternée de Schwarz*, C. R. Acad. Sci., Paris, Ser. A 286 (1978), pp. 325–328.

[203] R. LÖHNER AND K. MORGAN, *An unstructured multigrid method for elliptic problems*, Int. J. Numer. Methods Eng., 24 (1987), pp. 101–115.

[204] H. LÖTZBEYER AND U. RÜDE, *Patch-adaptive multilevel iteration*, BIT, 37 (1997), pp. 739–758.

[205] Y. MADAY, C. MAVRIPLIS, AND A. T. PATERA, *Nonconforming mortar element methods: Application to spectral discretizations*, in Domain decomposition methods 2, T. F. Chan, R. Glowinski, J. Périaux, and O. B. Widlund, eds., SIAM, Philadelphia, 1989, pp. 392–418.

[206] J. MANDEL, *Balancing domain decomposition*, Commun. Numer. Methods Eng., 9 (1993), pp. 233–241.

[207] A. M. MATSOKIN AND S. V. NEPOMNYASCHIKH, *The Schwarz alternating method in a subspace*, Sov. Math., 29 (1985), pp. 78–84.

[208] O. A. MCBRYAN, P. O. FREDERICKSON, J. LINDEN, A. SCHÜLLER, K. SOLCHENBACH, K. STÜBEN, C.-A. THOLE, AND U. TROTTENBERG, *Multigrid methods on parallel computers — a survey of recent developments*, IMPACT Comput. Sci. Engrg., 3 (1991), pp. 1–75.

[209] S. F. MCCORMICK, *Fast adaptive composite grid (FAC) methods: Theory for the variational case*, in Defect Correction Methods. Theory and Applications, K. Böhmer and H. J. Stetter, eds., Computing Suppl. 5, Springer, Vienna, 1984, pp. 115–121.

[210] ——, *Multilevel Adaptive Methods for Partial Differential Equations*, vol. 6 of Frontiers in Applied Mathematics, SIAM, Philadelphia, 1989.

[211] C. MESZTENYI, A. MILLER, AND W. SZYMCZAK, *FEARS: Details of mathematical formulation UNIVAC 1100*, Tech. Report BN-994, Univ. of Maryland, College Park, 1982.

[212] G. L. MILLER, S.-H. TENG, W. THURSTON, AND S. A. VAVASIS, *Geometric separators for finite-element meshes*, SIAM J. Sci. Comput., 19 (1998), pp. 364–386.

[213] W. F. MITCHELL, *A comparison of adaptive refinement techniques for elliptic problems*, ACM Trans. Math. Software, 15 (1989), pp. 326–347.

[214] ——, *A parallel multigrid method using the full domain partition*, Electron. Trans. Numer. Anal., 6 (1997), pp. 224–233. Special issue for proc. of the 8th Copper Mountain Conf. on Multigrid Methods.

[215] E. H. MOORE, *On certain crinkly curves*, Trans. American Math. Soc., 1 (1900),

pp. 72–90.

[216] M. MORAYNE, *On differentiability of Peano type functions. I, II*, Colloq. Math., 53 (1987), pp. 129–132, 133–135.

[217] D. MORGENSTERN, *Begründung des alternierenden Verfahrens durch Orthogonalprojektion*, Z. Angew. Math. Mech., 36 (1956), pp. 255–256.

[218] *Multi-grid methods I-VII*, Springer Berlin/ Birkhäuser, Basel/ Comput. Visual. Science, 1982/ 1986/ 1991/ 1993/ 1997/ 1999/ 2002.

[219] S. V. NEPOMNYASCHIKH, *Mesh theorems on traces, normalization of function traces and their inversion*, Sov. J. Numer. Anal. Math. Model., 6 (1991), pp. 223–242.

[220] ———, *Decomposition and fictitious domains methods for elliptic boundary value problems*, in Domain decomposition methods 5, T. F. Chan, D. E. Keyes, G. Meurant, J. S. Scroggs, and R. G. Voigt, eds., SIAM, Philadelphia, 1992, pp. 62–72.

[221] E. NETTO, *Beitrag zur Mannigfaltigkeitslehre*, Crelle J., 86 (1879), pp. 263–268.

[222] R. A. NICOLAIDES, *On the l^2 convergence of an algorithm for solving finite element equations*, Math. Comp., 31 (1977), pp. 892–906.

[223] R. NIEDERMEIER, K. REINHARDT, AND P. SANDERS, *Towards optimal locality in mesh-indexings*, in Proc. of the 11th Int. Symposium on Fundamentals of Computation Theory, no. 1279 in Lecture Notes in Computer Science, Springer, Berlin, Heidelberg, 1997, pp. 364–375.

[224] M. G. NORMAN AND P. MOSCATO, *The Euclidean traveling salesman problem and a space-filling curve*, Chaos Solitons Fractals, 6 (1995), pp. 389–397.

[225] J. T. ODEN, L. DEMKOWICZ, W. RACHOWICZ, AND T. A. WESTERMANN, *Toward a universal h-p adaptive finite element strategy. II: A posteriori error estimation.*, Comput. Methods Appl. Mech. Engrg., 77 (1989), pp. 113–180.

[226] J. T. ODEN, A. PATRA, AND Y. FENG, *Domain decomposition for adaptive hp finite element methods*, in Domain decomposition methods 7, D. E. Keyes and J. Xu, eds., vol. 180 of Contemp. Math., AMS, Providence, Rhode Island, 1994, pp. 295–301.

[227] W. F. OSGOOD, *A Jordan curve of positive area*, Trans. Amer. Math. Soc., 4 (1903), pp. 107–112.

[228] P. OSWALD, *On discrete norm estimates related to multilevel preconditioners in the finite element method*, in Constructive Theory of Functions, Proc. Int. Conf. Varna 1991, K. G. Ivanov, P. Petrushev, and B. Sendov, eds., Sofia, 1992, Bulg. Acad. Sci., pp. 203–214.

[229] ———, *On a BPX-preconditioner for P1 elements*, Computing, 51 (1993), pp. 125–133.

[230] ———, *Stable splittings of Sobolev spaces and applications*, Tech. Report Math/93/5, FSU Jena, 1993.

[231] ———, *Multilevel Finite Element Approximation*, Skripten zur Numerik, Teubner, Stuttgart, 1994.

[232] ———, *On the convergence rate of SOR: A worst case estimate*, Computing, 52 (1994), pp. 245–255.

[233] ———, *Best N-term approximation of singularity functions in two Haar bases*, tech. report, Bell Labs, Lucent Technologies, 1998.

[234] C.-W. OU, S. RANKA, AND G. FOX, *Fast and parallel mapping algorithms for irregular and adaptive problems*, in Proc. of Int. Conf. on Parallel and Distributed Systems, 1993.

[235] E. E. OVTCHINNIKOV AND L. S. XANTHIS, *Iterative subspace correction methods for*

thin elastic structures and Korn's type inequality in subspaces, Proc. R. Soc. Lond., Ser. A, 453 (1997), pp. 2003–2016.

[236] M. PARASHAR AND J. C. BROWNE, *Distributed dynamic data-structures for parallel adaptive mesh-refinement*, in Proc. of the Int. Conf. for High Performance Computing, 1995, pp. 22–27.

[237] ——, *On partitioning dynamic adaptive grid hierarchies*, in Proc. of the 29th Annual Hawai Int. Conf. on System Sciences, 1996, pp. 604–613.

[238] G. PEANO, *Sur une courbe qui remplit toute une aire plaine*, Mathematische Annalen, 36 (1890), pp. 157–160.

[239] P. PEISKER AND D. BRAESS, *A conjugate gradient method and a multigrid method for Morley's finite element approximation of the biharmonic equation*, Numer. Math., 50 (1987), pp. 567–586.

[240] A. PÉREZ, S. KAMATA, AND E. KAWAGUCHI, *Peano scanning of arbitrary size images*, in Proc. Int. Conf. Pattern Recognition, 1992, pp. 565–568.

[241] C. PFLAUM, *A multilevel algorithm for the solution of second order elliptic differential equations on sparse grids*, Numer. Math., 79 (1998), pp. 141–155.

[242] C. PFLAUM AND A. ZHOU, *Error analysis of the combination technique*, Numer. Math., 84 (1999), pp. 327–350.

[243] J. R. PILKINGTON AND S. B. BADEN, *Dynamic partitioning of non-uniform structured workloads with space-filling curves*, IEEE Trans. Parallel Distributed Systems, 7 (1996), pp. 288–300.

[244] A. POTHEN, *Graph partitioning algorithms with applications to scientific computing*, in Parallel Numerical Algorithms, D. E. Keyes, A. Sameh, and V. Venkatakrishnan, eds., Kluwer, Dordrecht, 1997, pp. 323–368.

[245] A. POTHEN, H. D. SIMON, AND K.-P. LIOU, *Partitioning sparse matrices with eigenvectors of graphs*, SIAM J. Matrix Anal. Appl., 11 (1990), pp. 430–452.

[246] W. H. PRESS, S. A. TEUKOLSKY, W. T. VETTERLING, AND B. P. FLANNERY, *Numerical recipes in C*, Cambridge Univ. Press, Cambridge, 2nd ed., 1992.

[247] J. S. PRZEMIENIECKI, *Theory of Matrix Structural Analysis*, McGraw-Hill, New York, 1968, ch. 9.

[248] D. QUINLAN, *Adaptive Mesh Refinement for Distributed Parallel Architectures*, PhD thesis, Dept. of Mathematics, Univ. of Colorado, Denver, 1993.

[249] R. RANNACHER, *Error control in finite element computations. An introduction to error estimation and mesh-size adaptation*, in Error control and adaptivity in scientific computing. Proc. of the NATO ASI, Antalya, H. Bulgak and C. Zenger, eds., vol. 536 of NATO ASI Ser., Ser. C, Math. Phys. Sci., Kluwer, Dordrecht, 1999, pp. 247–278.

[250] M. RENARDY AND W. ROGERS, *Introduction to Partial Differential Equations*, Springer, New York, 1993.

[251] A. REUSKEN, *Multigrid with matrix-dependent transfer operators for a singular perturbation problem*, Computing, 50 (1993), pp. 199–211.

[252] ——, *Convergence analysis of a multigrid method for convection-diffusion equations*, Numer. Math., 91 (2002), pp. 323–349.

[253] W. C. RHEINBOLDT AND C. K. MESZTENYI, *On a data structure for adaptive finite element mesh refinements*, ACM Trans. Math. Softw., 6 (1980), pp. 166–187.

[254] H. RITZDORF AND K. STÜBEN, *Adaptive multigrid on distributed memory computers*, in Multigrid methods IV. Proc. of the 4th European multigrid conf., Amsterdam,

vol. 116 of ISNM, Birkhäuser, Basel, 1994, pp. 77–95.

[255] M. C. RIVARA, *Algorithms for refining triangular grids suitable for adaptive and multigrid techniques*, Internat. J. Numer. Methods Engrg., 20 (1984), pp. 745–756.

[256] S. ROBERTS, S. KALYANASUNDARAM, M. CARDEW-HALL, AND W. CLARKE, *A key based parallel adaptive refinement technique for finite element methods*, in Proc. Computational Techniques and Applications: CTAC '97, B. J. Noye, M. D. Teubner, and A. W. Gill, eds., World Scientific, Singapore, 1998, pp. 577–584.

[257] U. RÜDE, *Data structures for multilevel adaptive methods and iterative solvers*, Tech. Report I-9217, TU München, Inst. für Informatik, 1992.

[258] ———, *Mathematical and Computational Techniques for Multilevel Adaptive Methods*, vol. 13 of Frontiers in Applied Mathematics, SIAM, Philadelphia, 1993.

[259] J. W. RUGE AND K. STÜBEN, *Algebraic multigrid (AMG)*, in Multigrid Methods, S. F. McCormick, ed., vol. 3 of Frontiers in Applied Mathematics, SIAM, Philadelphia, 1987, pp. 73–130.

[260] R. D. RUSSELL AND J. CHRISTIANSEN, *Adaptive mesh selection strategies for solving boundary value problems*, SIAM J. Numer. Anal., 15 (1978), pp. 59–80.

[261] H. SAGAN, *Space-Filling Curves*, Springer, New York, 1994.

[262] J. K. SALMON, M. S. WARREN, AND G. S. WINCKELMANS, *Fast parallel tree codes for gravitational and fluid dynamical N-body problems*, Int. J. Supercomputer Appl., 8 (1994), pp. 129–142.

[263] A. A. SAMARSKIJ, *Theorie der Differenzenverfahren*, Geest & Portig, Leipzig, 1984. translation of 'Teoriya raznostnykh skhem', Nauka, Moskva, 1977.

[264] H. SAMET, *The Design and Analysis of Spatial Data Structures*, Addison-Wesley, Reading, MA, 1990.

[265] G. SANDER AND B. F. DE VEUBEKE, *Upper and lower bounds to structural deformations by dual analysis in finite elements*, Tech. Report AFFDL-TR-66-199, Air Force Flight Dynamics Laboratory, Ohio, 1967.

[266] T. SCHIEKOFER, *Die Methode der Finiten Differenzen auf dünnen Gittern zur Lösung elliptischer und parabolischer partieller Differentialgleichungen*, Doktorarbeit, Universität Bonn, Inst. für Angew. Math., 1998.

[267] T. SCHIEKOFER AND G. ZUMBUSCH, *Software concepts of a sparse grid finite difference code*, in Proc. of the 14th GAMM-Seminar Kiel on Concepts of Numerical Software, W. Hackbusch and G. Wittum, eds., Notes on Numerical Fluid Mechanics, Vieweg, Braunschweig, 1999. submitted.

[268] R. SCHREIBER, *Scalability of sparse direct solvers*, in Graph Theory and Sparse Matrix Computations, Proc. of a workshop IMA program on "Applied linear algebra", A. George, J. R. Gilbert, and J. W.-H. Liu, eds., no. 56 in IMA Volume in Mathematics and its Applications, Springer, New York, 1993, pp. 191–209.

[269] C. SCHWAB AND R.-A. TODOR, *Sparse finite elements for elliptic problems with stochastic loading*, Numer. Math., (2003). electronically.

[270] H. A. SCHWARZ, *Über einige Abbildungsaufgaben*, Vierteljahresschrift Naturforsch. Ges. Zürich, 15 (1870), pp. 272–286.

[271] M. A. SCHWEITZER, *A Parallel Multilevel Partition of Unity Method for Elliptic Partial Differential Equations*, no. 29 in Lecture Notes in Computational Science and Engineering, Springer, Berlin, Heidelberg, 2003.

[272] M. A. SCHWEITZER, G. ZUMBUSCH, AND M. GRIEBEL, *Parnass2: A cluster of*

dual-processor PCs, in Proc. of the 2nd Workshop Cluster-Computing, Karlsruhe, W. Rehm and T. Ungerer, eds., no. CSR-99-02 in Chemnitzer Informatik Berichte, TU Chemnitz, TU Chemnitz, 1999, pp. 45–54.

[273] G. H. SHORTLEY AND R. WELLER, *The numerical solution of Laplace's equation*, J. Appl. Phys., 9 (1938), pp. 334–344.

[274] W. SIERPIŃSKI, *Sur une novelle courbe continue qui remplit toute une aire plaine*, Bull. Acad. Sci. de Cracovie (Sci. math. et nat., Série A), (1912), pp. 462–478.

[275] H. D. SIMON AND S.-H. TENG, *How good is recursive bisection?*, SIAM J. Sci. Comput., 18 (1997), pp. 1436–1445.

[276] B. F. SMITH, *Domain decomposition algorithms for the partial differential equations of linear elasticity*, PhD thesis, Dept. Computer Science, Courant Institute, New York, 1990.

[277] ——, *A domain decomposition algorithm for elliptic problems in three dimensions*, Numer. Math., 60 (1991), pp. 219–234.

[278] B. F. SMITH, P. E. BJØRSTAD, AND W. D. GROPP, *Domain Decomposition: Parallel Multilevel Methods for Elliptic Partial Differential Equations*, Cambridge Univ. Press, Cambridge, 1996.

[279] B. F. SMITH AND O. B. WIDLUND, *A domain decomposition algorithm using a hierarchical basis*, SIAM J. Sci. Statist. Comput., 11 (1990), pp. 1212–1220.

[280] S. A. SMOLYAK, *Quadrature and interpolation formulas for tensor products of certain classes of functions*, Dokl. Akad. Nauk SSSR, 148 (1963), pp. 1042–1045.

[281] M. SNIR, S. OTTO, S. HUSS-LEDERMAN, D. WALKER, AND J. DONGARRA, *MPI: The Complete Reference*, vol. 1, MIT Press, Cambridge, MA, 2nd ed., 1998.

[282] S. L. SOBOLEV, *L'algorithme de Schwarz dans la theorie de l'elasticit*, C. R. (Dokl.) Acad. Sci. URSS, n. Ser., (1936), pp. 243–246.

[283] K. SOLCHENBACH, C. A. THOLE, AND U. TROTTENBERG, *Parallel multigrid methods: Implementation on SUPRENUM-like architectures and applications*, in Supercomputing, vol. 297 of Lecture Notes in Computer Science, Springer, Berlin, Heidelberg, 1987, pp. 28–42.

[284] D. A. SPIELMAN AND S.-H. TENG, *Spectral partitioning works: Planar graphs and finite element meshes*, in Proc. 37th Annual IEEE Symposium on Foundations of Computer Science, Los Alamitos, CA, 1996, IEEE Computer Society Press, pp. 96–105.

[285] L. STALS, *Parallel Implementation of Multigrid Methods*, PhD thesis, Australian National Univ., Dept. of Mathematics, 1995.

[286] L. STALS AND U. RÜDE, *Techniques for improving the data locality of iterative methods*, Tech. Report MRR97-038, School of Math. Sciences, Australian National Univ., 1997.

[287] R. P. STEVENSON, *Robustness of the additive and multiplicative frequency decomposition multi-level method*, Computing, 54 (1995), pp. 331–346.

[288] Q. F. STOUT, *Topological matching*, in Proc. 15th ACM Symp. on Theory of Computing, 1983, pp. 24–31.

[289] K. STÜBEN AND U. TROTTENBERG, *Multigrid methods: Fundamental algorithms, model problem analysis and applications*, in Multigrid Methods, Proc. Conf., Köln-Porz, W. Hackbusch and U. Trottenberg, eds., vol. 960 of Lecture Notes in Mathematics, Springer, Berlin, Heidelberg, 1982, pp. 1–176.

[290] X.-C. TAI AND M. ESPEDAL, *Rate of convergence of some space decomposition methods for linear and nonlinear problems*, SIAM J. Numer. Anal., 35 (1998), pp. 1558–1570.

[291] X.-C. TAI AND J. XU, *Global convergence of subspace correction methods for convex optimization problems*, Math. Comp., (2001). electronically.

[292] V. N. TEMLYAKOV, *On the approximation of periodic functions of several variables with bounded mixed difference*, Sov. Math. Dokl., 22 (1980), pp. 131–135.

[293] ——, *Approximation of functions with bounded mixed derivative*, Proc. of the Steklov Institute of Mathematics, 178 (1989), p. 121.

[294] C. H. TONG, T. F. CHAN, AND C. C. J. KUO, *A domain decomposition preconditioner based on a change to a multilevel nodal basis*, SIAM J. Sci. Statist. Comput., 12 (1991), pp. 1486–1495.

[295] U. TROTTENBERG, C. OOSTERLEE, AND A. SCHÜLLER, *Multigrid*, Academic Press, San Diego, CA, 2000.

[296] R. A. VAN DE GEIJN, *Massively parallel LINPACK benchmark on the Intel Touchstone Delta and iPSC(r)/860 systems*, in 1991 Annual Users' Conf. Proc., Dallas, TX, 1991, Intel Supercomputer Users' Group.

[297] P. VANĚK, J. MANDEL, AND M. BREZINA, *Algebraic multigrid by smoothed aggregation for second and fourth order problems*, Computing, 56 (1996), pp. 179–196.

[298] L. VELHO AND J. GOMES, *Digital halftoning with space filling curves*, Computer Graphics, 25 (1991), pp. 81–90. SIGGRAPH'91.

[299] R. VERFÜRTH, *A Review of A Posteriori Error Estimation and Adaptive Mesh-Refinement Techniques*, J. Wiley & Teubner, Chichester, 1996.

[300] D. VOORHIES, *Space-filling curves and a measure of coherence*, in Graphics Gems II, J. Arvo, ed., Academic Press, San Diego, CA, 1994, pp. 26–30.

[301] C. H. WALSHAW AND M. BERZINS, *Dynamic load-balancing for PDE solvers on adaptive unstructured meshes*, Concurrency: Practice and Experience, 7 (1995), pp. 17–28.

[302] M. S. WARREN AND J. K. SALMON, *A portable parallel particle program*, Comput. Phys. Commun., 87 (1995), pp. 266–290.

[303] R. E. WEBBER AND Y. ZHANG, *Space diffusion: An improved parallel halftoning technique using space-filling curves*, in Proc. ACM Comput. Graphics Ann. Conf. Series, 1993, p. 305ff.

[304] P. WESSELING, *The rate of convergence of a multiple grid method*, in Numerical Analysis, Proc. 8th bienn. Conf., Dundee, G. A. Watson, ed., vol. 773 of Lecture Notes in Mathematics, Springer, Berlin, Heidelberg, 1980, pp. 164–180.

[305] ——, *An Introduction to Multigrid Methods*, J. Wiley, Chichester, 1992.

[306] O. B. WIDLUND, *Some Schwarz methods for symmetric and nonsymmetric elliptic problems*, in Domain decomposition methods 5, T. F. Chan, D. E. Keyes, G. A. Meurant, J. S. Scroggs, and R. G. Voigt, eds., SIAM, Philadelphia, 1992, pp. 19–36.

[307] R. D. WILLIAMS, *Performance of dynamic load-balancing algorithm for unstructured mesh calculations*, Concurrency: Practice and Experience, 3 (1991), pp. 457–481.

[308] ——, *Voxel databases: A paradigm for parallelism with spatial structure*, Concurrency: Practice and Experience, 4 (1992), pp. 619–636.

[309] G. WITTUM, *On the robustness of ILU smoothing*, SIAM J. Sci. Statist. Comput., 10 (1989), pp. 699–717.

[310] J. XU, *Theory of Multilevel Methods*, PhD thesis, Cornell Univ., 1989.

[311] ——, *Iterative methods by space decomposition and subspace correction*, SIAM Rev., 34 (1992), pp. 581–613.

[312] ——, *Two-grid discretization techniques for linear and nonlinear PDEs*, SIAM J. Numer. Anal., 33 (1996), pp. 1759–1777.

[313] J. XU AND J. ZOU, *Some nonoverlapping domain decomposition methods*, SIAM Rev., 40 (1998), pp. 857–914.

[314] H. YSERENTANT, *On the multi-level splitting of finite element spaces*, Numer. Math., 49 (1986), pp. 379–412.

[315] ——, *Old and new convergence proofs for multigrid methods*, in Acta Numerica, A. Iserles, ed., vol. 2, Cambridge Univ. Press, Cambridge, 1993, pp. 285–326.

[316] P. M. D. ZEEUW, *Matrix-dependent prolongations and restrictions in a blackbox multigrid solver*, J. Comput. Appl. Math., 33 (1990), pp. 1–27.

[317] C. ZENGER, *Sparse grids*, in Parallel Algorithms for Partial Differential Equations, Proc. 6th GAMM-Semin., Kiel, W. Hackbusch, ed., no. 31 in Notes on Numerical Fluid Mechanics, Vieweg, Braunschweig, 1991, pp. 241–251.

[318] X. ZHANG, *Multilevel Schwarz methods*, Numer. Math., 63 (1992), pp. 521–539.

[319] ——, *Multilevel Schwarz methods for the biharmonic Dirichlet problem*, SIAM J. Sci. Comput., 15 (1994), pp. 621–644.

[320] ——, *Analysis of additive multilevel methods*, East-West J. Numer. Math., 8 (2000), pp. 71–82.

[321] G. ZUMBUSCH, *Adaptive parallele Multilevel-Methoden zur Lösung elliptischer Randwertprobleme*, SFB-Report 342/19/91A, TUM-I9127, SFB 342, TU München, 1991.

[322] ——, *Simultanous h-p Adaptation in Multilevel Finite Elements*, Doktorarbeit, FU Berlin, 1995.

[323] ——, *A sparse grid PDE solver; discretization, adaptivity, software design and parallelization.*, in Advances in Software Tools in Scientific Computing, H. P. Langtangen, A. M. Bruaset, and E. Quak, eds., no. 10 in Lecture Notes in Computational Science and Engineering, Springer, Berlin, Heidelberg, 2000, ch. 4, pp. 133–177.

[324] ——, *On the quality of space-filling curve induced partitions*, Z. Angew. Math. Mech., Suppl., 81 (2001), pp. 25–28.

[325] D. ZUNG, *Some approximation characteristics of the classes of smooth functions of several variables in the metric of* $\mathcal{L}_2$, Russ. Math. Surv., 34 (1979), pp. 161–162. alternative spelling 'Dinh Dung'.

[326] ——, *Number of integral points in a certain set and the approximation of functions of several variables*, Math. Notes, 36 (1984), pp. 736–744.

Index